SCIENCE AND RELIGION IN SEVENTEENTH-CENTURY ENGLAND

RICHARD S. WESTFALL

SCIENCE AND RELIGION IN SEVENTEENTH-CENTURY ENGLAND

Ann Arbor Paperbacks

THE UNIVERSITY OF MICHIGAN PRESS

First edition as an Ann Arbor Paperback 1973

ISBN 0-472-06190-9
Published in the United States of America by
The University of Michigan Press and simultaneously
in Don Mills, Canada, by Longman Canada Limited
Manufactured in the United States of America

TO

Gloria

Preface

LIKE most authors who sit down to acknowledge their debts, I am overcome by the near infinity of obligations and must limit myself to mentioning the small minority which, like the visible part of an iceberg, only attests to the greater bulk unseen. Most of my studies were carried on in four libraries—the Yale Library and the Huntington Library in the United States and the British Museum and the Library of the Royal Society in London; without the cooperation of their staffs I could not have completed the work. I owe particular obligations to the Royal Society for permission to use the Boyle Papers and other MSS, to the British Museum for the use of a large number of MSS in their collection, to the provost and fellows of King's College, Cambridge, for permission to use their collection of Newtonian material, the Keynes MSS, and to the Bodleian Library for access to the Locke MSS. To the State University of Iowa I express my thanks for financial aid toward the work's publication.

Many people have given me the benefit of their criticism and advice. Frederick L. Kilgour of the Yale Medical Library and Hartley Simpson, Dean of the Yale Graduate School, contributed their knowledge of the history of science and of 17th-century England. The advice of S. T. Bindoff of the Institute of Historical Research in London was equally valuable. I could not allow this book to appear without singling out three persons whose aid I cannot repay but can mention with appreciation. The study was conceived originally under the inspiration of Franklin L. Baumer of the Yale History Department and was carried out under his guidance and counsel. He has been all that a student could wish for in a teacher—and more than he deserves. My mother performed many functions from criticizing the text to proofreading the copy. Finally, my wife can honorably take her place among other wives who have aided and sustained their mates through similar labors.

R. S. W.

Grinnell, Iowa

Preface to the Paperback Edition

Twenty years have passed since the period, 1950-52, when I did most of the research on which *Science and Religion in Seventeenth Century England* is based. In 1950, the history of science scarcely existed as an academic discipline; at no point during my education at a leading university did I have the opportunity to take a course devoted to it. During the intervening years, our knowledge of the history of science has grown beyond measure, with the scientific revolution of the sixteenth and seventeenth centuries as the foremost object of attention; and during the same years, responding to the stimulus of my own work, I have shaped my career in terms of the new discipline. Perhaps I am simply too rigid to subject a position once taken to serious reexamination, but as I look back from my present vantage point, it still appears to me that I correctly defined one of the basic questions that the rise of modern science presented to the intellectual history of the Western world. In 1600, Western civilization found its focus in the Christian religion; by 1700, modern natural science had displaced religion from its central position. Every step forward in the understanding of seventeenth-century science, far from blurring that crucial contrast, has served only to heighten it. The relation of science to religion in the seventeenth century does seem to me now, as it did twenty years ago, the central question in the history of modern Western thought.

When I recall the untroubled confidence with which I set out to reform historical study, I am amazed that I had the good sense (well prodded by good advice, to be sure) to confine my attention to a sharply limited segment of the vast question. If, as I like to hope, the intervening years have brought deeper understanding of the scientific revolution, they have also diluted the self-assurance, and I am no more ready now than I was then to tackle the question on a less confined front. Were I to start again now on this specific study, I would leave its broad outlines unchanged. Certainly I would not venture into the morass of Puritanism and science, which seems to exist only to swallow all those who attempt to cross it. Although the question was muddy in the early 1950's, at least it was relatively fresh. Twenty years of polemics appear to me to have done nothing to clarify it; they have done even less to keep it fresh. At any rate, the debate has concentrated on the causation of modern science; for my question it could only be a digression. I now think I was mistaken to omit Hobbes from the study. Whatever his differences from the other virtuosi, his version of the mechanical philosophy was the one original system of nature contributed by an English philosopher before Newton. The book ought to include him in a central role. Certainly I would alter the discussion of Newton, the study of whose papers has been one of the major accomplishments of the history of science during the present generation. As I now realize, Newton's religious thought, with its Hermetic coloring, was considerably more complicated than what I presented in the final chapter. For all its complication, however, it does appear to have led to substantially the same conclusion. To the eighteenth century, which did not have access to his private papers, Newton's stance could appear only one step short of Deism. If his papers reveal a different posture, their God has become the active source of the laws of motion, so identical to the mechanical order that a mere change of vocabulary could suffice to eliminate him. More even than I realized, Newton represents the modifications of the Christian tradition as it adjusted itself to the new guiding force in Western thought.

R. S. W.

Bloomington, Indiana

Contents

CHAPTER 1

The Problem

It is, indeed, marvelous that science should ever have revived amid the fearful obstacles theologians cast in her way.

Lecky, *History of the Rise and Influence of the Spirit of Rationalism in Europe*

For more than half a century in England there was among Christians a degree of friendliness often amounting to an enthusiastic welcome for the New Philosophy.

Raven, *Natural Religion and Christian Theology*

A GREAT WATERSHED lying across the history of Western civilization, the 17th century marks the beginning of the distinctively modern world. New institutions, new conventions, new concepts are apparent in many fields.

Politically, the territorial state triumphed over feudal decentralization in western Europe; and in England royal absolutism dissolved into Parliamentary sovereignty. The migration of power and energy away from the Mediterranean basin, a movement apparent as early as the 16th century, was confirmed and made definite; when the 17th century closed, France and England, twin colossi of the West, stood astride the Continent. Economically, the age witnessed the consolidation of capitalism. And not least important, the 17th century saw the growth to maturity of modern natural science. The "great instauration," which Sir Francis Bacon announced with the dawn of the century, achieved in the century's twilight a glorious culmination in the work of Sir Isaac Newton. The 17th century, the "age of genius" in Whitehead's phrase, laid the foundation upon which modern culture builds.

No element of European civilization revealed the changing patterns of life and thought more clearly than the Christian religion. For a thousand years the core of European culture, it now began to experience a decline of influence. The triumphant advance of secularization proceeded on all fronts. Economic activity shook off the guiding hand of Christian ethics and declared itself to be an independent aspect of life, governed by its own impersonal laws. The state emancipated itself from religious tutelage and the church became its servant. The religious fury of the Thirty Years' War dominated the first half of the century, but the Thirty Years' War was the last religious war. Although Europe was willing after 1648 to fight for political security and economic advantage, the passion that ignited religious conflagrations had burned too low to set another blaze. During the century England progressed from the Puritans' stern battle against the Arminians to the opening phase of the controversy over deism, which was acrimonious enough, to be sure, but was conducted, significantly, with the pen instead of the sword. Religious temperatures abated across western Europe. By the end of the century England had accepted religious toleration, if not religious freedom, as a necessary condition of internal repose. Although Europe was still almost unanimously Christian, although the Christian religion was virtually unchallenged openly, Christianity was the focus of European life and culture no longer.

At the same time that Christianity was losing influence, it was undergoing other changes, especially during the latter half of the century. A new intellectual current, the achievements of natural science, were raising questions that could not be ignored. There was the possibility that investigation of nature might so absorb a man that he would neglect the worship of God, and investigations were particularly to be feared if the discovery of second causes called the existence or power of the First Cause into doubt. Natural science rested on the concept of natural order, and the line that separated the concepts of natural order and material determinism was not inviolable. The mechanical idea of nature, which accompanied the rise of modern science in the 17th century, contradicted the assertion of miracles and questioned the reality of divine providence. Science, moreover, contained its own criteria of truth, which not only repudiated the primacy of ancient philosophers but also implied doubt as to the Bible's authority and regarded the attitude of faith enjoined by the Christian religion with suspicion. Confronted with such problems, a man was not faced, to be sure, with an exclusive choice—either natural science or Christianity: every one of the problems could be resolved in a variety of ways to reconcile science with religion. But the mere fact of reconciliation meant some change from the pattern of traditional Christianity. With the growing prestige of science—it achieved immense prestige after the publication of Newton's *Principia*—its reconciliation with Christianity came more and more to mean the adjustment of Christian beliefs to conform to the conclusions of science. The rise of natural science to a position of intellectual dominance over Christianity in the 17th century was therefore a major factor in the change that the Christian religion was undergoing.

Science was only one element in a complex of forces operating to change traditional Christianity. Although no single country could fully reflect a process which was not confined within political boundaries, England presented a typical picture of most of the factors at work. The great Civil War and the multiplicity of conflicting sects which were spawned during the Interregnum left a deep impression on the minds of thoughtful men and led them to question traditional assumptions. Many were convinced that religious strife bade fair to destroy not only England but Christianity itself. Some basis of mutual agreement and toleration had to be found, and the demand set men in search of common rational fundamentals of religion on which all Christians could unite. Even before the Civil War the spectacle of Christian contending against Christian had stimulated an irenical movement which tried to affirm the fundamentals held in common by the different Christian, or at least Protestant, denominations. Its roots in England stretched back to the Elizabethan age and governmental policy at that time. Immediately before the Civil War, William Chillingworth published its most profound theological justification, *Religion of Protestants* (1638), while on the practical level John Durie labored untiringly for unity. A different aspect of the movement, which approached the problem of religious peace through skepticism, appeared in the natural religion of Lord Herbert of Cherbury. Even when posed by a superficial thinker, Lord Herbert's inquiries probed basic questions, and his natural religion was pregnant with the possibility of future development. Following the Civil War the irenical tradition with its emphasis on fundamental beliefs exerted great influence on English religious thought.

Another result of the Civil War helped to spur the quest for religious fundamentals. When the dike of ecclesiastical authority had been cast down in 1642, a torrent of religious radicalism had swept through the breach, flooding the land. The extravagance of some left-wing sectarians, "enthusiasts" as they were called, presented an alarming spectacle to reasonable men, and confronted them with a pressing choice. Either they must renounce all religion as an imposture and a delusion, or they must demonstrate the rational foundations of religion to justify their faith. For all but a handful the first alternative was out of the question, and hatred of enthusiasm drove reasonable men to search out the rational foundations on which religion rests.

"Rational foundations" is another name for natural religion. Since the days of the medieval Schoolmen the employment of natural religion in defense of Christianity had been an established tradition. That the importance generally assigned to it increased markedly during the 17th century was due not only to sectarian madness but also to England's, and Europe's, expanding geographical outlook. When exploration brought Christendom into contact with a multitude of hitherto unknown pagan peoples, western Europe's first impulse was to convert them; but they also led some men to question the universality and binding force of the Christian revelation if many peoples had never heard of Christ. This consideration was one of the determinants of Lord Herbert's natural religion, and others followed his steps. Both John Locke and Isaac Newton, for example, asked whether a heathen who has never heard of Christ can be saved; both answered in effect that God proportions His demands on men to their conditions, and that the principles of natural religion unaided by revelation can lead the well-meaning heathen to eternal life.

England's Protestant heritage also contained seeds which were able to grow into new patterns of religion in the genial climate of the Restoration. With the fires of religious war burning down and the need for organizational strength and rigid conformity slackening, the implications of the Protestant emphasis on individual responsibility could be probed. It could lead to frank speculations and inquiries into the meaning of the new natural science for Christianity and into the other religious problems of the day, which were difficult at best for the faithful Catholic. For example, John Locke's vigorous belief in the individual's sole responsibility for his own salvation accompanied radically unorthodox speculations, which sought to adjust Christianity to the social and intellectual conditions of the late 17th century. Moreover, the practical result of the Reformation in splitting ecclesiastical authority was enhanced in England by the Civil War. The obstacles hampering the individual's quest for religious truth were destroyed; effective authority to enforce conformity and to maintain the status quo no longer existed.

In addition, specific elements of Protestantism conformed to the viewpoint of natural science. The general Protestant contempt for medieval Scholasticism was reflected throughout the writings of the English scientists. It helped to prepare the nearly unanimous and surprisingly facile acceptance which the mechanical conception of nature met among the English scientists, and the mechanical conception in turn imposed the necessity of accommodation upon Christian beliefs. Furthermore, the Calvinist God in His remote majesty resembles the watchmaker God of the mechanical universe, suggesting that the Calvinist tenor of English theology helped to make the mechanical hypothesis congenial to English scientists. Particularly in the sermons of John Wallis, the most rigid Calvinist among the English scientists, the image of God as arbitrary will laying down eternal and unchangeable laws for the creation seems ready to merge with Robert Boyle's portrait of the Divine Mechanic constructing His inexorable machine. Perhaps no Christian belief collided more immediately with the mechanical conception of nature than the belief in miracles; once again Protestant teachings helped to prepare the rationalization of miracles accepted by most of the English scientists. In accordance with the Protestant rejection of transubstantiation and contempt for Catholic "superstition," and in accordance with the necessity to repudiate miracles alleged in support of Catholicism, Protestantism generally refused to accept the validity of miracles that occurred after biblical times. Most of the English scientists were willing to accept the biblical miracles, although the miracles contradicted scientific principles; but granting those exceptions alone, they insisted that nature is otherwise a mechanically determined order. Without doing violence to their Protestant convictions, they were able to hold a position on miracles which was not in flagrant contradiction to their scientific theories.

The influence of Calvinist teachings can be seen throughout the English scientists' attempts to reconcile the doctrine of providence with the mechanical order of nature. The potential contradiction between the doctrine of providence and the mechanical conception of nature appeared in Dr. Samuel Clarke's epistolary controversy with Leibniz early in the 18th century. Leibniz' fundamental argument was that God created the universe because He saw that it was rational and good. In Leibniz' view nature is a wholly comprehensible order spread open before human understanding; a meaningful notion of divine intervention could not be reconciled with this opinion. Dr. Clarke's reply embodied the position stated by most of the English scientists who considered the question of providence. Contrary to Leibniz' rationalism it adopted a positivistic attitude, declaring that the universe is rational and good only because God created it. In agreeing that the universe is a rational order, this position provided an intellectual sanction for scientific investigation. On the other hand it opened the way for the affirmation of divine providence by denying that the universe conforms entirely to human reason. In good Calvinist fashion it asserted that man cannot fully penetrate divine reason, which is a rule unto itself—that God has ends beyond human comprehension. Calvinism and Catholicism were not diametrically opposed in this argument by any means. Calvin's emphasis on God's irrationality found its source among the Scholastic nominalists, and Thomism never asserted that man can comprehend all of God's ends. Yet the frequent reiteration by English scientists, as they strove to reconcile providence with the mechanical order, that God has ends beyond our understanding suggests the relatively greater emphasis in Protestantism on human inability to probe the divine nature.

Too great a role should not be attributed to Protestantism in determining the course of religious development in England through its influence on natural science in the late 17th century. That modern science did not appear exclusively in Protestant countries is a fact too obvious to need stress. Galileo, Descartes, and Pascal, among other scientific geniuses, were the products of Catholic culture. On the other hand Protestant England was the undeniable leader in natural science at the end of the 17th century, and Catholics were prominent neither in numbers nor in accomplishments among the English scientists. Irene Parker, Dorothy Stimson, Richard F. Jones, George Rosen, and Robert K. Merton have argued with some cogency the connection between Puritanism—vague and ill-defined a movement though it was—and modern science.[1] The influence of Protestantism on natural science is nebulous and difficult to determine. It can be suggested, however, that Protestantism provided an atmosphere more conducive to scientific investigation as such than was Catholicism; perhaps also Protestantism was more conducive to the acceptance of the peculiar mechanical conception of nature which accompanied early modern science and to its reconciliation with religious beliefs.

On a practical plane below the level of theoretical concepts Protestantism exerted a different influence on the English religious scene during the latter half of the 17th century. From its beginning Protestantism, and especially Calvinism, embodied a fuller acceptance of the world than Catholicism. It rejected the monastic ideal that the highest religious life is realized by withdrawing from the world. While warning its followers not to be soiled by worldly values, Protestantism required them to accept

1. Irene Parker, *Dissenting Academies in England* (Cambridge, England, 1914), mentions the connection of Puritanism with science. Dorothy Stimson offers some statistical evidence for it in an article, "Puritanism and the New Philosophy in 17th Century England," *Bulletin of the* [*Institute of the*] *History of Medicine, 3* (1935), 321–34. Richard F. Jones amplifies and documents it in *Ancients and Moderns* (St. Louis, 1936) and in *The Triumph of the English Language* (Stanford, 1953). George Rosen attempts to extend the connection to the sectarians, in "Left-wing Puritanism and Science," *Bulletin of the History of Medicine, 15* (1944), 375–80. By far the most intensive treatment is Robert K. Merton's book-length article, "Science, Technology, and Society in Seventeenth Century England," *Osiris, 4* (1938), 360–632. Although I believe that some connection between Puritanism and early modern science has been established, the definitive treatment of it remains to be written.

the world as the Lord's vineyard in which they were placed to labor. Calvinism in particular evolved a dynamic social ideal which enjoined the Christian to employ existing institutions as instruments to show forth the glory of God. The Christian's labor in his calling became a spiritual duty as well as a material necessity; the Christian's calling was the peculiar sphere in which he realized Christian values. After 1660, when the Puritan New Jerusalem had collapsed and when English economic life was expanding, the Calvinist social ideal degenerated toward crass materialism, which equated worldly success with divine blessing. In this form it harmonized with and abetted the dominant tone of Restoration life. England was enjoying prosperity such as she had never known before. Unworldliness was not at a premium. The Baconian ideal, the utilitarianism, of the scientists of the period reflected the prevailing attitude. John Beale, a clergyman from Herefordshire, published tracts on the profits to be gained by the scientific cultivation of orchards; and Nehemiah Grew composed a treatise on the means of increasing England's wealth by the application of scientific knowledge. Both men were sincere and practicing Protestants, but like many Englishmen of the Restoration period they intended to taste the fruits of Bacon's Kingdom of Man while they waited for the eternal but distant repast in the Kingdom of Heaven. In a word, the values of this world were challenging the values of the other world. Men still believed that salvation was the ultimate human goal, but they did not expect salvation to preclude reasonable gratification here and now. The worldliness of Restoration England was bound to affect the tenor of religious thinking. It was reflected in a growing tendency among preachers and among authors of devotional books to slacken the rigors of Christian living, to reduce religious practice to reasonable moral virtue, and to equate moral virtue with the bourgeois values of diligence, frugality, moderation, and respectability.

A new and growing feeling of confidence also pervaded English thought—confidence in human capacities, and confidence in the possibilities of life. It appeared in many contexts, and its implications for Christian thought were profound. In the verbal controversy over the relative attainments of ancients and moderns, opinion during the early years of the century generally awarded the laurels to the ancients. The decision was reversed by the end of the century as confident men assured themselves that the modern world had exceeded antiquity. A new spirit informed interpretations of the biblical prophecies. Where earlier the prophecies were thought to predict the coming of Antichrist and the end of the world, they were now seen to point toward a future millennium when a new and better life would arise from expanding knowledge. John Locke's famous treatise on civil government was based on an optimistic conception of man which could not be reconciled with the traditional Christian doctrines of original sin and human depravity.

The late 17th century was an age of questioning when traditional values and ideas, especially Christian concepts, were being tested and often repudiated. The irenical movement's search for fundamentals encouraged a growing emphasis on natural religion which arose from other causes. As a result both of the nature of Protestantism itself and of the collapse of ecclesiastical compulsion in England, bold and even radical speculations on religious questions were possible as never before. Certain Protestant beliefs were facilitating the acceptance of modern natural science which would in time impose changes on Christian beliefs. To these factors were added the worldliness of Restoration England and the feeling of confidence that pervaded the age. A note of uncertainty among the champions of Christianity indicates that they perceived, however vaguely, that they labored in altered circumstances. Although Christianity stood virtually unchallenged by any other religion, although it enjoyed the support of political authority, Christianity was on the defensive. Instead of preaching its message with confidence, it was seeking to justify itself. Its apologists tried to prove that it was perfectly rational and therefore suitable for reasonable men. A provision in Robert Boyle's will, much applauded at the time for its pious intent, was a revealing straw in the wind. To bequeath money for religious purposes was a practice sanctioned by long tradition; to endow a series of lectures in defense of religion, however, was a type of bequest that would not have been made before uncertainty and doubt were well advanced.

Into this atmosphere the added factor of modern natural science was injected. The mechanical conception of nature, which was the first synthesis of modern science into an elaborated theory of the whole of creation, embodied a radical break with the idea of nature that for centuries had been intimately bound up with Christian theology. When the mechanical theory rapidly gained acceptance among educated circles, specifying the terms in which men envisaged the universe and life, a major transformation in Western thinking had taken place. That the new view of the universe and of man's place in it would have affected Christian theology profoundly, whatever the other influences on theology at the time, appears evident from historical precedents. Christianity had always stated its doctrines in terms of prevailing philosophical principles. Augustine's works had preached Christianity in the language of neoplatonism; and when Aristotle was rediscovered, the Schoolmen of the 13th century synthesized Christianity and Aristotelianism. Now in the 17th century a new and widely accepted theory of the cosmos, of man, and of knowledge—a theory which contradicted specific Christian beliefs, questioned others, and embodied the possibility of materialism—could not be kept within a separate compartment. Christianity was forced to take notice of the new natural science. But it is idle to speculate on the impact modern science would have had on religion had other factors not been present. Historical development is an organic process, the parts of which do not subsist in isolation. Many of the other factors influencing Christian thought in the late 17th century also helped to foster the growth of natural science. That science did affect Christianity so profoundly was due in part to the fact that science reached maturity in an age when orthodoxy was shaken and Christian thought in flux. Because old beliefs were unsettled and natural science offered a new criterion of certainty, Christianity felt the impact of science through its whole frame.

This study endeavors to illuminate some aspects of the interaction of science and religion in the late 17th century by studying the opinions of the scientists (the "virtuosi," as they were called). Its view is restricted to the English virtuosi, an arbitrary limitation to be sure, but one justified to some extent by the leadership that England held in the development of science at the time. In the minds of the English virtuosi the historian can find a microcosmic picture of Christianity and natural science interacting upon each other. It is necessary to say "interacting," because modern science was produced by a Christian society, and however much science may have affected Christianity, it could not itself remain untouched by the religious beliefs of the men who conceived it. Christians of patent sincerity, with but few exceptions, the virtuosi saw in the discoveries of natural science confirmation of their religious beliefs. In the many books that they published to show how God has revealed Himself in nature they effected a satisfying consolidation of science and religion. So strong was the influence of Christianity upon them that it helped to mold their conception of nature, softening the harsh mechanical outline with the comforting light of divine benevolence. At the same time the influence of natural science was helping to change their idea of Christianity profoundly, leading them to emphasize rational or demonstrable elements at the expense of suprarational mysteries, until by the end of the century some virtuosi were professing a form of Christianity which scarcely differed from the Enlightenment's religion of reason. The virtuosi present a concrete example of the relations of science and religion in the 17th century, with each intellectual system influencing the other. In a broader sense the study of Christianity and the virtuosi is a case history of intellectual change, an example of the process through which mankind lays aside a pervasive world view which has governed its intellectual outlook and takes up another.

In the second half of the 19th century, when modern historical scholarship first took up the relations of natural science and religion in the 17th century, it concentrated on the conflicts of theology with the conclusions of science. W. E. H. Lecky, John W. Draper, and Andrew D. White wrote, naturally, under the influence of their own age.[2] Darwinism had made its stormy entrance into Western thought, causing the intellectual world to

2. W. E. H. Lecky, *History of the Rise and Influence of the Spirit of Rationalism in Europe,* 2 vols. London, 1865. John W. Draper, *History of the Conflict between Religion and Science,* New York, 1875. Andrew D. White, *A History of the Warfare of Science with Theology in Christendom,* 2 vols. New York, 1896 (enlarged version of *Warfare of Science,* first published 1876). None of the three discusses the 17th century exclusively. With Lecky it is the major center of interest, but his idea of rationalism includes a great deal more than natural science. See my Bibliography, below, for a fuller discussion of the works mentioned in this paragraph.

ring with the conflict of science and religion; and the three historians tended to read the spirit of the late 19th century into the earlier period. Although the spirit of contemporary scholarship has changed markedly from the confident temper of the 19th century, the early studies, particularly those of Lecky and White, contain much of value. Paul Hazard's recent investigation of the origins of the Enlightenment adopts their point of view insofar as it treats of natural science.[3] For the most part, however, the 20th century's revival of Christianity has ushered in a new attitude. An important school of research, represented by E. A. Burtt, R. F. Jones, and Basil Willey, has continued to emphasize the incompatibility of 17th-century science with Christianity, but it has shifted concern from the welfare of science to the welfare of religion.[4] Tending to ignore the immediate quarrels of science with theology, this school devotes its attention to the ultimate implications that modern science holds for religion. Another group of studies by Alfred North Whitehead, R. G. Collingwood, and Charles E. Raven suggests that Christianity influenced the concepts of early modern science deeply, whatever the incompatibility between the two.[5] The problem of science and religion in the 17th century has also been approached through the opinions of individual virtuosi, most successfully in Mitchell S. Fisher's discussion of Robert Boyle and in Canon Raven's biography of John Ray.[6] Still lacking, however, is a complete investigation of the relations of early modern science and religion, emphasizing at once their conflicts and the reconciliation between them that was effected.

Paul H. Kocher has recently brought out an extensive study of science and religion in Elizabethan England which undertakes

3. *La Crise de la conscience Européenne*, 3 vols. Paris, 1934.

4. E. A. Burtt, *The Metaphysical Foundations of Modern Physical Science*, London, 1925. Jones, *Ancients and Moderns*. Basil Willey, *Seventeenth Century Background*, London, 1934.

5. Alfred North Whitehead, *Science and the Modern World*, New York, 1925. R. G. Collingwood, *The Idea of Nature*, Oxford, 1945. Charles E. Raven, *Natural Religion and Christian Theology*, Cambridge, England, 1953.

6. Mitchell S. Fisher, *Robert Boyle, Devout Naturalist*, Philadelphia, 1945. Charles E. Raven, *John Ray, Naturalist*, Cambridge, England, Cambridge University Press, 1942.

to cover the opinions of both the scientists and the divines.[7] Since the Elizabethan age did not witness a dramatic change of opinion on the questions involved, Kocher is able to treat the relations of science and religion more as a static problem than a study of the 17th century will allow. He focuses his attention primarily on the area where science and religion came most directly into contact—the extent of determinism in the natural order. The investigation of religion and science in the 17th century must take note of a dynamic situation in which opinions change a great deal as the century proceeds; it must examine aspects of religion further removed from the central area of friction. Since the virtuosi were at once deeply committed to Christianity and thoroughly familiar with natural science, the relations of science and religion can perhaps best be studied through their opinions; comprehension of the virtuosi should contribute to an understanding of the age as a whole.

In the article entitled "The English Virtuoso in the Seventeenth Century" Walter E. Houghton maintains that in the usage of the period "virtuoso" meant a dilletante who titillated his sensibilities by dallying with natural rarities and curiosities.[8] He argues that a virtuoso was not a true scientist but a gentleman of leisure, half-heartedly stifling the boredom of life by esoteric studies, the figure ridiculed in Shadwell's satire *The Virtuoso.* Houghton's definition does not correspond entirely to 17th-century usage. Although the term was employed in his sense, a more common usage applied it to the true scientists. The word came into use in the middle of the century; its first instance listed in the Oxford English Dictionary occurred in Brent's translation of Sarpi's *Council of Trent,* published in 1651. According to the same dictionary the 17th century used "virtuoso" to mean "one who has a general interest in arts and sciences or who pursues special investigations in one or more of these; a learned person; a scientist, savant, or scholar." The second meaning of the word (Houghton's definition)—a student or collector of rarities and natural curiosities—became current about the same time; the Oxford Dictionary's first instance

7. *Science and Religion in Elizabethan England,* San Marino, California, 1953.
8. Walter E. Houghton, Jr., "The English Virtuoso in the Seventeenth Century," *Journal of the History of Ideas,* 3 (1942), 51–73, 190–219.

is found in a book by John Evelyn in 1662. The only other meaning (the one that is common today), namely, a skilled musical performer, did not begin until the 18th century. This study employs "virtuoso" in the sense of the first definition—that is, one who has a general interest in arts and sciences, or who pursues special investigations in one or more of these; the meaning is further restricted to those interested specifically in natural science. The virtuosi themselves defined the term in their books and letters. Samuel Hartlib wrote to John Worthington about the first formal meetings of the group that later called itself the Royal Society, saying that "there is a meeting every week of the prime Virtuosi." [9] Joseph Glanvill's "Address to the Royal Society," published in his *Scepsis Scientifica*,[10] spoke of the "virtuosi and enquiring spirits of Europe." The word is used extensively in Henry Stubbe's attacks on Sprat and Glanvill, and is evidently restricted in meaning to members of the Royal Society. Robert Boyle employed it repeatedly, including it even in the title of one of his more famous books, *The Christian Virtuoso;* he defined it, in the plural, as "those that understand and cultivate experimental philosophy." [11] The virtuosi, then, were those who took an active interest in promoting the growth of natural science.

Since virtuosi in the 17th century were an even more indefinite group than are scientists today, it is difficult to decide what men fit the definition. The criteria here used are experimental work, publication of books, correspondence with other virtuosi, or participation in the activities of the Royal Society. A virtuoso might answer to other names as well, of course; many of the leading ones, for instance, were practicing clergymen. In an age when science had not become an institutionalized activity, few men could devote their full time to it. Even the first Astronomer Royal, John Flamsteed, had to supplement his income by holding a church living.

9. Hartlib to Worthington, December 17, 1660; *The Diary and Correspondence of Dr. John Worthington,* ed. James Crossley (2 vols. Manchester, 1847–84), *1,* 346. I have modernized spelling and punctuation of 17th-century source material.

10. *Scepsis Scientifica: or, Confessed Ignorance the Way to Science; in an Essay of the Vanity of Dogmatizing and Confident Opinion,* London, 1665.

11. *The Works of the Honourable Robert Boyle,* ed. Thomas Birch (6 vols. London, 1772), *5,* 513–14.

The virtuosi here studied worked during roughly the latter half of the 17th century. Sir Kenelm Digby, John Wilkins, and Samuel Hartlib were active in natural philosophy during the decade of 1640, although they continued their work into the Restoration period. Isaac Newton serves as the terminal figure even though two of his younger contemporaries, Edmond Halley and John Flamsteed, are included.

To sketch a portrait of the average virtuoso would be impossible. A heterogeneous group, the virtuosi belonged to no one class, pursued sundry occupations, varied in wealth from riches to poverty, and differed in religious affiliations. Only their interest in natural philosophy bound them together. But that one bond was compelling: seeking one another's company, the virtuosi formed groups which mark the first hesitant efforts of science to organize the institutions of a social activity.

During the early years of modern science the unpaid but devoted labor of focal correspondents supplied the lack of formal organizations in providing an element of cohesion among scattered investigators. Samuel Hartlib (1600?–62), the German born immigrant and Baconian enthusiast, was the first of the notable correspondents in England. Although Hartlib was not an experimenter himself, he maintained a broad web of contacts, solicited pamphlets on useful knowledge from his friends, and published them for the benefit of society—in all he assembled a mountain of information. His pet scheme, an office of "public address," as he called it, foresaw the organization of scientific endeavor on a national scale. Having exhausted his fortune and his health in promoting the growth of science, he died in the year that the Royal Society received its first charter. Among Hartlib's most faithful correspondents was John Beale (1603–83?), a clergyman in Herefordshire and Somersetshire. Beale's pamphlet *Herefordshire Orchards* (1656), which Hartlib published, epitomized his ardent embrace of the Baconian ideal. A common interest in scientific knowledge brought both Hartlib and Beale into close association with Robert Boyle, the wealthy son of the Earl of Cork, from whom they seem to have differed on most other counts.

Hartlib's death in 1662 coincided with the appearance of the man who took up his function of correspondence. Henry Oldenburg (1615?–77), like Hartlib, was born on the Continent. Facility in languages, together with untiring energy, made him indispensable as secretary of the Royal Society, and because of his correspondence with Continental scientists, the Royal Society acquired a European reputation quite beyond its early accomplishments.

Yoked with Oldenburg in the harness of the Royal Society was his antithesis and enemy, Robert Hooke (1635–1703). The stormy petrel of the virtuosi, Hooke was a natural genius who lacked the stability to follow any idea to its conclusion. Introduced to scientific work as Boyle's assistant, he probably discovered the law of pressure and volume in gases that bears Boyle's name. Not only did he possess the instinct of a mechanical genius which enabled him to perfect the air pump and invent the universal joint, but he roamed at will over the entire continent of science, producing a brilliant suggestion wherever he stepped. Inevitably Oldenburg's methodical plodding clashed with Hooke's mercurial genius. They hated each other bitterly; Hooke accused Oldenburg of stealing his discoveries, and Oldenburg assiduously fanned the controversies between Hooke and Newton. Yet somehow they worked together, and virtually unassisted they kept the Royal Society alive. Oldenburg provided the administration, and Hooke contributed most of the substance to the early meetings. Over them Lord William Brouncker (1620?–84), aristocratic courtier and governmental official, presided as president, contributing the experience of another background and personality to the progress of science.

Among the other members of the society diversity was as great. Sir Kenelm Digby (1603–65), a courtier and a Catholic, turned from a profligate life to the solace of philosophy upon the death of his beautiful wife, Venetia. Always a lover of esoteric knowledge, he practiced astrology and expounded the curative virtue of "sympathetic powder," until he was termed the Pliny of his day for lying. He followed more solid studies as well, and he joined as a founding member of the Royal Society with Robert Boyle (1627–91), a devout Protestant who rejected a title, a bishopric, and the presidency of the society in order to remain simply the Honorable Robert Boyle, investigating nature in peace. No early member of the society enjoyed a greater reputation. From the time when his house in Oxford was a congregating point for the virtuosi during the Commonwealth until his death in 1691 he was a nucleus around whom others grouped. As a philosopher of science, Boyle systematized and justified the mechanical conception of nature held by most of the virtuosi.

Medical doctors were prominent among the group, both in numbers and in eminence. Walter Charleton (1619–1707), an extreme Royalist and a high Anglican, was active during the first years of the Royal Society and he later served for a time as president of the Royal College of Physicians. Nehemiah Grew (1641–1712) came from a family of dissenting Protestants. While managing a London practice, he found time to conduct important research on the anatomy of plants. One of the most prominent physicians in London, Richard Lower (1631–91), was a pioneering physiologist who did much to explain the heart and the circulatory system. Another physician, John Mapletoft (1631–1721), rose through a London practice to the Gresham chair of physic, from which he retired when nearly fifty years old to enter the ministry. William Petty (1623–87), who practiced medicine successfully for a time, finally devoted most of his energy to public service as an official. A universal genius, Petty touched on many fields of natural philosophy and eventually helped to found the science of statistics and the modern study of economics. Two doctors among the virtuosi dealt with in this study did not belong to the Royal Society. They were not wholly isolated from the other virtuosi, however. Thomas Sydenham (1624–89), the pioneer clinical observer, associated closely with both Mapletoft and Locke. Sir Thomas Browne (1605–82), who is famous primarily as a literary figure, carried on a desultory correspondence with Oldenburg from his home in Norfolk and was connected distantly to the Royal Society through his son, Dr. Edward Browne, F.R.S.

Clerics were nearly as prominent as doctors. That they were is not surprising; most of the virtuosi were deeply religious, and clergymen found little difficulty in reconciling the ministry with interest in natural philosophy. John Wilkins (1614–72), who survived his active affiliation with the Puritan party and his marriage to Cromwell's sister to become Bishop of Chester, was one of the organizing spirits of the scientific movement. As Warden of Wadham College during the Commonwealth, he made the college a center of scientific activity. Many of the young men who congregated there later participated in the founding of the Royal Society. John Wallis (1616–1703), from whose letters we know of the "Invisible College" of 1645, also identified himself with the Puritan cause during the Civil War but accepted the Restoration and became a royal chaplain. A bellicose character engaged in endless quarrels and controversies, Wallis was also a great mathematician and a keen student of mechanics.

Not all of the clerical virtuosi had close Puritan affiliations. The most serene and attractive personality among them, Isaac Barrow (1630–77), refused to compromise his Anglican convictions throughout the Commonwealth. Lucasian professor of mathematics at Cambridge and the equal of Wallis in ability, Barrow taught Newton mathematics, helped to prepare the ground for the latter's discovery of the calculus, and then resigned his chair in order to devote his time fully to the Church.

The two most influential publicists of the Royal Society were also staunchly Anglican churchmen. Thomas Sprat (1635–1713) had a pliable nature. Recognizing that the road to success did not run through Puritanism after the Restoration, he became prominent among high church circles and attained the bishopric of Rochester. Sprat's only service to the cause of science was the famous *History of the Royal Society*. As publicity it was a significant contribution, but evidently he had to be coerced into finishing it.[12] While Joseph Glanvill (1636–80) remained in a more modest position as rector of Bath, his service to natural science was greater. As the author of *Plus Ultra, Philosophia Pia,* and other essays, he was the most skillful apologist of the virtuosi. The great astronomer John Flamsteed (1646–1719) was also a clergyman, holding a living in Surrey and serving it periodically between his indefatigable observations. In nearly half a century as Astronomer Royal

12. A letter from John Wallis to Sir Robert Moray, October 26, 1664, in the files of the Royal Society (Guard Book *W 1,* fol. 12), indicates that Sprat was neglecting and delaying the completion of the *History*, to the annoyance of the members of the Royal Society. Wallis suggested that influence with the Duke of Buckingham be used to force Sprat to finish the book. Wallis considered its publication essential to the interests of the society.

he won a reputation as an observer second only to Tycho Brahe.

John Ray (1627–1705), the leading naturalist of his day, led a relatively secluded life of study following his retirement from Cambridge when the Act of Uniformity was imposed in 1662. From his home in Essex he turned out a prodigious volume of work, pioneering studies of botany, zoology, ornithology, ichthyology, and entomology. In spite of his isolation he maintained close contact with the Royal Society and with its leading members. The most renowned of English architects, Sir Christopher Wren (1632–1723), demonstrated genius in science that might well have led him to an eminence equal to Newton's had not the great fire of London diverted his energy to building. John Locke (1632–1704) combined his philosophical speculations and his interest in science with an active public life which extended from his affiliation with Lord Ashley in 1667 to his resignation from the Council of Trade in 1700. Edmond Halley (1656–1742), on the other hand, devoted his life almost entirely to science, ultimately gaining the Savilian chair in geometry, despite rumors of religious skepticism which had earlier denied him the chair in astronomy. He later succeeded Flamsteed in the Royal Observatory.

The life of the greatest virtuoso, Isaac Newton (1642–1727), was a mixture of contrasting elements. For over thirty years he lived as a secluded scholar deeply absorbed in intellectual pursuits. Deadly serious and reported to have laughed only once, he was capable of concentrating so intensely on a problem that he forgot his meal on a tray beside him. From this period of his life stem all of his scientific discoveries. A few years following the publication of the *Principia* he suddenly began to solicit a public position; and when he received the wardenship of the Mint, he gladly abandoned active work in science for the life of London. Although Newton's creative period was at an end, he served for many years as president of the Royal Society, contributing his personality and reputation to the group which from its formation had been the center around which the English virtuosi gathered.

In all, the virtuosi were a fair cross-section of educated English society in the late 17th century. Far from being isolated scholars in an ivory tower, they were profoundly involved in the life of their society—in the government, in the universities, in the church, in the professions. Among their other pursuits they found time and inclination to take up the study of science. In the 17th century their interest in natural science was beginning to distinguish them as a group—not a group separated from the rest of society, but a group who accepted this new and growing interest as an important part of their lives, a group who sought out one another to discuss the problems of natural philosophy. Nothing could better indicate the natural bond uniting the virtuosi than the unplanned gathering to hear Christopher Wren's lecture at Gresham College late in 1660. The virtuosi present on that day decided to formalize their meetings and to organize a society for the discussion of natural philosophy. Thus there came into being the organization which named itself the Royal Society when Charles II granted it a charter. In the 17th century the Royal Society was not the exclusive body that it is today. Anyone interested, however vaguely, in natural science could become a member. Birch's *History of the Royal Society*, mostly a transcript of the minutes through 1687, records no instance of a candidate's being rejected. Since its principal function in the 17th century was to provide a place for the discussion of natural philosophy, the survival of the Royal Society indicates the bond of common interest which made the virtuosi a living group.

Perhaps it can be added that Christianity was another recognized bond among the virtuosi. Thomas Hobbes, who was notorious in their eyes as an atheist, did not apply and was not suggested for membership in the Royal Society. Hobbes is not dealt with directly in this study, although strict application of the definition of virtuoso here used would include him as readily as some of the others. Hobbes was singular in his outlook, however, and completely opposed in religious views to the virtuosi. Since he would require a volume by himself, he has been excluded. His exclusion is further justified by his isolation from the virtuosi during his life. In a sense, all of the virtuosi's religious works can be looked upon as one long disapproving footnote to Hobbes' philosophy. Had he been proposed for membership in the Royal Society, he would probably have been the one exception to their practice of admitting anyone who claimed interest in natural philosophy.

In shunning the company of Hobbes, the virtuosi were expressing their opinion that science did not challenge Christianity. But the virtuosi were not the only Christians in England. If they refused to believe that their work threatened their religion, other Christians who did regard science as a grave danger were ready to set them right. To be sure, the men who chose to see the new science as a threat were only a small minority. Nevertheless, substantial attacks were mounted upon science in the name of Christianity. If they did not stay its victorious advance, they did exert an important influence on the minds of the virtuosi by forcing them to consider the religious implications of their work and by posing questions which the virtuosi had to answer. In a pamphlet cleverly entitled *The New Planet No Planet: or the Earth No Wandering Star; except in the Wandering Heads of Galileans* [13] Alexander Ross, the vigorous defender of traditional orthodoxy, fell upon John Wilkins's *Discourse Concerning a New Planet*. His denunciation of Copernican astronomy amounted to a condemnation of all scientific investigation. "You say it's but a novelty in philosophy, but I say it intrenches upon divinity," he protested hotly; "for divinity tells us that the standing of the sun and moving of the earth are the miraculous works of God's supernatural power; your new philosophy tells us that they are the ordinary works of nature" (p. 2). For all his extravagant diction Ross had put his finger on the central issue—the relative authority of natural philosophy and inspired divinity. As he developed his virulent argument, he continually stressed the presumption of science in seeking to override the authority of God and to judge the causes of things by human reason:

> Whereas you say that astronomy serves to confirm the truth of the Holy Scripture you are very preposterous; for you will have the truth of Scripture confirmed by astronomy, but you will not have the truth of astronomy confirmed by Scripture; sure one would think that astronomical truths had more need of Scripture confirmation than the Scripture of them.[14]

Richard Baxter, a man of greater moderation, took issue with the same aspect of science in *The Arrogancy of Reason against*

13. London, 1646.
14. Ibid., p. 117.

Divine Revelations Repressed (1655). Without assailing any specific scientific theory, Baxter directed his protest more toward the scientific spirit than toward the body of knowledge proposed by science. Some men, he reported, will not believe the truth of a revelation because they cannot understand how the thing revealed is caused—questioning God's decrees of predestination, for instance, because the causes are incomprehensible, denying the very work of creation since they do not know how it was done. When Baxter inquired into the origins of the unhappy distemper of infidelity, as he called it, he decided that men are naturally desirous of knowledge, seeking to know things through their evidence. They are therefore impatient to be admitted into the presence chamber of truth and to see her naked without delay. No arguments, no authority will suffice; men must see for themselves.

> If the wisest men in the world tell them that they see it or know it; if the workers of miracles, Christ and His Apostles, tell them that they see it; if God Himself tells them that He sees it; yet all this does not satisfy them unless they may see it themselves. . . . Every man has an understanding of his own, and therefore would have a sight of the evidence himself, and so have a nearer knowledge of the thing, and not only a knowledge of the truth of the thing by the testimony of another, how infallible soever.[15]

In a word, Richard Baxter feared that the very temper of scientific research endangered the foundations of Christianity.

Other facets of science also worried Baxter. In *Reasons of the Christian Religion* (1667) he spoke out against the Epicurean philosophy which was in vogue among the virtuosi and which was carried to an extreme by Thomas Hobbes. The Epicureans, Baxter declared, looked at corporeal things so much that they overlooked the noblest aspects of nature; since they studied nothing but matter and motion thoroughly, they reduced everything to those principles. "And like idle boys who tear out all the hard leaves of their books and say they have learned all when they have learned

15. *The Arrogancy of Reason against Divine Revelations Repressed* (bound with *The Unreasonableness of Infidelity*) (London, 1655), Pt. IV, p. 39.

the rest, so do they cut off and deny the noblest parts of nature and then sweep together the dust of agitated atoms and tell us that they have resolved all the phenomena in nature."[16]

Meric Casaubon found the same dangers in Epicurean atomism in two treatises, *Of Credulity and Incredulity in Things Divine and Spiritual* (1670) and *Of Credulity and Incredulity in Things Natural and Civil* (1672). Devotion to the study of second causes, he feared, might lead men to forget the supremacy of God; examination of the sensible qualities of bodies might end in the denial of everything that is not perceptible to the senses. Thomas Baker's moderate appraisal of the relative authority of science and religion, *Reflections upon Learning* (1700), concluded with the same apprehension—that the study of nature would distract the attention of men away from the eternal truths of religion.

Perhaps the most bitter denunciation of the virtuosi was delivered by Henry Stubbe, who subjected them to a withering fire of ridicule and condemnation in a series of pamphlets issued in 1670. The sincerity of Stubbe can no longer be accepted, since evidence that he was a hired character assassin has been uncovered.[17] But the work of a clever and unscrupulous rabble rouser, playing upon the secret fears of people, may reveal the apprehensions of more sincere men. Stubbe's attack on Glanvill, *The Plus Ultra Reduced to a Nonplus,* contained every possible suggestion that the work of the Royal Society was detrimental to

16. *The Reasons of the Christian Religion* (London, 1667), p. 498.

17. Harcourt Brown, *Scientific Organizations in Seventeenth Century France* (Baltimore, 1934), pp. 255–7. A MS life of Dr. Baldwin Hȧmey, a leader of the Royal College of Physicians in the 17th century, written by his nephew and now lodged in the library of the R.C.P., says that Hamey, fearing that the Royal Society would infringe upon the sphere of the R.C.P., hired Stubbe to attack the young organization. I have found a letter from John Wallis to Henry Oldenburg, dated October 25, 1670, the year of Stubbe's pamphlets, which reported that Dr. Pierce, president of Magdalen College, Oxford, sent Stubbe a piece of plate worth five or six pounds "for his good service" (Royal Society, Guard Book *W 1,* fol. 113). A series of letters from Stubbe to Hobbes, written in 1656–57 when Hobbes was carrying on a pamphlet war with Wallis, further reveals Stubbe's character. Stubbe wrote as an intermediary for the Independent faction in Oxford, urging Hobbes on to the attack, secretly furnishing him with scandal, telling Hobbes at the same time that he would have to disavow him in public (British Museum, Add. MS 32,553, fols. 5–34). In all, Stubbe appears as a clever but wholly venal scoundrel.

religion. He asserted that the Royal Society was destroying the weapons with which Christianity had been defended by repudiating the old philosophy and Scholastic divinity. Nor would Stubbe allow the virtuosi's contention that the study of material things prepares the mind readily to acknowledge immaterial beings and gladly to praise God for the richness of creation. So long as their studies were restricted to material bodies, he asserted, the virtuosi could never furnish grounds for belief in the immaterial. The concept of a mechanical geometric universe threatened the primacy of God in the creation. If the Lord be regulated by the rules of geometry and mechanical motion in the government of the world, he declared, "I cannot any way comprehend how God can do any miracles." [18] Stubbe's *Censure upon Certain Passages Contained in the History of the Royal Society* took issue with several of Sprat's statements about religion, including the assertion common among the virtuosi that the learned praises of one who has studied the Almighty's works are more acceptable to God than the blind wonder of the ignorant. With St. Paul he replied that any work done without the inspiration of grace is worthless. No matter how much and how well an experimental philosopher studies the creation, he will not thereby become more acceptable to God than a man who studies the Scripture with humility and reverence and seeks to be accepted through the merit of Christ. A "Psalm of David," Stubbe declared, "the Te Deum, the Magnificat, in a blind and ignorant but devout Christian, will be better accepted than a Cartesian anthem." [19] In all, Stubbe earned his wages well, and in doing so he managed—probably by accident, when one considers his cynical nature—to uncover some of the deep religious questions that natural science provoked.

The dangers that the champions of Christianity found in science comprise essentially two main points. On the one hand science promoted intellectual arrogance which led a man to prefer his own notions to the inspired word of God. On the other hand the mechanical atomic philosophy was likely to end in pure material-

18. *The Plus Ultra Reduced to a Nonplus* (London, 1670), pp. 172–3.

19. *A Censure upon Certain Passages Contained in the History of the Royal Society, as Being Destructive to the Established Religion and Church of England* (London, 1670), p. 38.

ism. The virtuosi were well aware of the charges brought against them; numerous references in their works revealed their concern to show that their studies were really directed to the welfare of Christianity. To understand their replies it is necessary to examine their science and their religion as a whole.

CHAPTER 2

The Harmony of Science and Religion

> The discoveries that a naturalist makes of the wisdom of God may be resembled to those that a traveller makes of water in a ship; where the further he descends down a great river, the more water he discovers . . . ; till arriving at the sea, he sees water much further than he did before, and yet sees cause to conclude that he might discover much further, if his eyes were better, or assisted with telescopes.
>
> Boyle, *Boyle Papers*

> I confess it is an easy thing to lose oneself in such speculations about God's wisdom and His other attributes displayed in the framing and conduct of His creatures.
>
> Boyle, *The Christian Virtuoso*

TO THE CHARGE of irreligion the virtuosi replied with a display of reverence for the works of God which both filled their books and reflected one of their fundamental traits, for the virtuosi were pious men. It is a revealing commentary on them that wherever their studies led them, they found themselves following the footprints of God. "The heavens declare the glory of God; and the firmament showeth His handiwork." The words had been sung centuries before, to be sure, and every age since had repeated them; yet probably no age had sung them with quite the fervor of the late 17th century. The period had the privilege of revolutionary discovery in so many fields; it pioneered on so many fronts. No age before had understood how well the heavens do declare the glory of God, and no age since has known the untutored surprise of first uncovering the hidden beauty of nature. Robert Hooke, for example, was the first man in history to examine a flea under a microscope. In his *Micrographia* he published a huge and horrendous reproduction of it—and described the appalling monster as "beautiful." Familiarity with microscopic structures had not blunted the virtuosi's sense of wonder. Wherever they turned in the study of nature, they discovered new wonders which provoked ever new adoration for their Creator. Nothing was more true of the virtuosi's own experience than their repeated assertion that the study of nature only leads the soul more surely to God. John Beale, one of the Royal Society's more enthusiastic members, spoke for them all when he mentioned "the lawful and religious delight which should result from beholding the curious and wonderful frame of this our visible world." [1]

For the virtuosi the study of nature and the perception of her beauty were positive religious experiences. To them the world was not a meaningless turning of gears and wheels; it was an order instinct with the intelligence of the Creator, a mighty testimony of His grandeur, worthy to display His glory. Nature's declaration of the Creator's glory inspired them with a feeling of awed surprise which runs as a theme throughout their works. It appears typically in Henry Oldenburg's defense of natural philosophy. Our "*Christian* philosophers," he asserted to those who condemned the virtuosi, using the word "Christian" to contrast the virtuosi to the pagan Aristotle, will "acquaint you with the true works and wonderful contrivances of the Supreme Author." [2] John Beale referred to the "wonderful frame" of the visible world; Oldenburg spoke of the "wonderful contrivances" of the Supreme Author. The choice of words was not an accident. Wherever they turned, the virtuosi found a wonderful frame and wonderful contrivances, so that every investigation was also an act of worship.

Three hundred years have passed, but the spontaneity of the virtuosi can still be recaptured in part. Chance phrases and exclamations scattered through their works make it possible to catch fleeting glimpses of nature through their eyes and to experience something of their spirit of wonder and natural reverence. Nehemiah Grew described the microscopic structure of plants in the vocabulary of artistic criticism. The "staple of the stuff is so exquisitely fine that no silkworm is able to draw anything near so small a thread. So that one who walks about with the meanest stick holds a piece of nature's handicraft which far surpasses the

1. Beale to Boyle, October 8, 1670; Royal Society, Guard Book *B 1*, fol. 55.
2. Henry Oldenburg, "Preface to the Third Year," *Philosophical Transactions*, *2* (1666), 413.

most elaborate woof or needlework in the world."[3] The sheer immensity of the heavens excited the imagination of young Christopher Wren. "Who can better magnify the arm that expanded the heavens," he demanded, "than he who tells you that seven thousand miles will fall short of the diameter of this earth, and yet that this diameter repeated a thousand times will not reach the sun, or this distance between the sun and us repeated a thousand times reach the nearest fixed star."[4] John Flamsteed finished an astronomical calculation and awed by the magnificence of the heavens dashed off the reflex conclusion "Sit Deo cuncta laus et gloria."[5] Even the reputed skeptic Edmond Halley forgot his skepticism in composing some verses to prefix the first edition of Newton's *Principia.*

> Behold the regions of the heavens surveyed,
> And this fair system in the balance weighed!
> Behold the law, which (when in ruin hurled
> God out of chaos called the beauteous world)
> Th' Almighty fixed, when all things good He saw!
> Behold the chaste, inviolable law![6]

No effort was required for men who had been reared in the teachings of Christianity to weave their science into their religion by referring the glory of creation to the Creator. Christianity taught that the world was created by God *ex nihilo;* the virtuosi were discovering a creation more worthy of His excellence even than Christians had earlier suspected. Glory they had found, and it must be God's.

Above all else, the virtuosi were not indifferent to the object they studied, or coldly detached. Robert Boyle often used the simile of a watchmaker examining the work of the most highly

3. Nehemiah Grew, "Epistle Dedicatory" to chap. 2 of *Anatomy of Plants,* London, 1682.

4. Christopher Wren, "Inaugural Oration as Gresham Professor of Astronomy," in James Elmes, *Memoirs of the Life and Works of Sir Christopher Wren* (2 vols. London, 1823), 2, appendix 44.

5. Cited in Francis Bailey, *An Account of the Revd John Flamsteed, the First Astronomer-royal* (London, 1835), preface, p. xxii n.

6. Translation from the Latin, printed in *Correspondence and Papers of Edmond Halley,* ed. Eugene F. MacPike (Oxford, the Clarendon Press, 1932), p. 207.

skilled craftsman of his trade: how well such a workman could appreciate the labor that had gone into a masterpiece, and with what enthusiasm he could admire it. The virtuosi were the watchmakers admiring the skill of a master, not mere tourists surveying it placidly, museum-goers dutifully absorbing a veneer of culture, but watchmakers whose inner life responded to the genius embedded in the work. Robert Hooke, for example, was not a devout man; yet when he looked through his microscope, the first man, as he realized, to observe the minute structure of many things, the beauties that he uncovered inspired him as much as a sunset ever did Wordsworth. Hooke's *Micrographia* has recorded his response to the newly discovered world. Starting with inorganic nature, he described, among other things, the delightful figures shown by grains of sand. He included a tiny shell because it offered a commentary upon nature as a whole. "For by it we have a very good instance of the curiosity of nature in another kind of animals which are removed by reason of their minuteness beyond the reach of our eyes; so that as there are several sorts of insects, as mites and others, so small as not yet to have had any names . . . and small fishes, as leeches in vinegar, and small vegetables, as moss and rose-leaf-plants, and small mushrooms, as mould; so are there, it seems, small shellfish likewise, nature showing her curiosity in every tribe of animals, vegetables, and minerals."[7] In the vegetable kingdom the size of moss gave food for contemplation.

> I know not whether all the contrivances and mechanisms requisite to a perfect vegetable may not be crowded into an exceedingly less room than this of moss . . . and I have already given you the description of a plant growing on rose leaves that is abundantly smaller than moss, insomuch that near 1000 of them would hardly make the bigness of one single plant of moss. And by comparing the bulk of moss with the bulk of the biggest kind of vegetable we meet with in story (. . . in some hotter climates, as Guinea and Brazil . . .) we shall find that the bulk of the one will exceed the

7. Robert Hooke, *Micrographia,* Vol. 13 of R. W. T. Gunther, *Early Science in Oxford* (14 vols. London and Oxford, 1920–45), p. 80.

> bulk of the other no less than 2985984 millions, or 2,985,984, 000,000; and supposing the production on a rose leaf to be a plant, we shall have of those Indian plants to exceed a production of the same vegetable kingdom no less than 1000 times the former number; so prodigiously various are the works of the Creator, and so all-sufficient is He to perform what to man would seem impossible, they being both alike easy to Him, even as one day and a thousand years are to Him as one and the same time.[8]

Nature reached her peak in the animal kingdom. The structure of a feather, for instance, seemed a triumph of ingenuity. Whereas strong bodies are usually dense and heavy, the quill combines strength with lightness in a hollow tube filled with bubbles; "and therefore should nature have made a body so broad and so strong as a feather almost any other way than what it has taken, the gravity of it must necessarily have many times exceeded this." Feathers are so made that they weave together naturally to form a mat strong enough to hold firmly together despite their violent beating against the air.

> And it argues an admirable providence of nature in the contrivance and fabric of them [he observed]; for their texture is such that though by any external injury the parts of them are violently disjoined . . . yet for the most part of themselves they readily rejoin and recontext themselves and are easily . . . settled and woven into their former and natural posture; for there are such an infinite company of those small fibres in the under side of the leaves and most of them have such little crooks at their ends that they readily catch and hold the stalks they touch.[9]

Hooke discovered the multiple eyes of the fly—and remarked the wisdom that had provided a creature unable to turn its head with the means of seeing in several directions. He investigated the generation of insects—and marveled at the goodness of providence which had bestowed upon insects instinctive knowledge of the

8. Ibid., pp. 134–5.
9. Ibid., pp. 165–7.

best place to lay their eggs. Insects as a whole excited his aesthetic sense. "The strength and beauty of this small creature," he said of the flea, "had it no other relation at all to man, would deserve a description." The feather-winged moth "afforded a lovely object both to the naked eye and through a microscope." The blue fly "is a very beautiful creature and has many things about it very notable." "To conclude," he referred to the gnat, "take this creature altogether, and for beauty and curious contrivances it may be compared with the largest animal upon the earth. Nor does the All-wise Creator seem to have shown less care and providence in the fabric of it than in those which seem most considerable." [10] In his Cutlerian lectures on light Hooke paused to admire the eye, the structure of which is so curiously contrived "that it is beyond the wit of man to imagine anything could have been more complete. Nay, it could never have entered into the imagination or thought of man to conceive how such a sensation as vision could be performed had not the All-wise Contriver of the world endued him with the faculty and organ of seeing itself." [11] Hooke's words are the vocabulary of natural awe. And they echo the response of all of the virtuosi. " 'Tis the contemplation of the wonderful order, law, and power of that we call nature that does most magnify the beauty and excellency of the divine providence, which has so disposed, ordered, adapted, and empowered each part so to operate as to produce the wonderful effects which we see; I say wonderful because every natural production may be truly said to be a wonder or miracle if duly considered." [12] What a volume of meaning is compressed into the one sentence. Pursuing science but finding miracles, the virtuosi responded emotionally as well as intellectually to the object of their studies.

Religiously inspired as they were, they naturally, even unconsciously, concluded that the discoveries of natural philosophy would not contradict the teachings of Christianity which they believed in and practiced. They neither thought that past discoveries challenged Christian truths, nor feared the future progress

10. Ibid., pp. 182, 195, 210.

11. Robert Hooke, *The Posthumous Works of Robert Hooke,* ed. Richard Waller (London, 1705), p. 121.

12. Ibid., pp. 423–4.

of knowledge to which they looked forward. Without holding their conclusions up as a higher truth by which the validity of Christianity was to be judged, the virtuosi thought of themselves as simple investigators of God's creation, studying the natural revelation as theologians studied the Scriptural revelation. Their faith in the harmony of science and religion expressed itself concretely in the life of the Baconian reformer Samuel Hartlib. Hartlib was not a scientist himself, but he took an intelligent interest in the progress of science and welcomed the formation of the Royal Society, although he was too sick with his final illness to become a member himself.[13] As one who had drunk deeply at the fountain of Comenius' universal philosophy, Pansophy, Hartlib believed in the harmony of truth and devoted his life to its advancement on all planes, technological, scientific, and religious. When he was a young man, he had engaged in a society called Antilia, which was seeking to found a Christian utopia devoted to the pursuit of truth. Although the original scheme for Antilia had died, it suddenly revived in 1660 just as the Royal Society was taking form. Lord Skytte, a Swedish nobleman, promised support to the utopian project, and for a time it seemed that Antilia might be translated into the world of reality. As Hartlib hopefully discussed the plans in his letters to John Worthington, he could not separate the success of Antilia from the formation of the Royal Society.

13. On Hartlib's absence from the roll of the Royal Society there has been some comment implying that his interests did not lie in the studies they pursued. The letters from his last years, however, show a decided interest in the meetings of the virtuosi. It should be remembered that the Royal Society—it did not even have that name before Hartlib's death—was not then an organization with the great prestige it has now. One joined to discuss natural philosophy with other interested men. It is true that during the first flush of enthusiasm following its formation, the society enjoyed a brief spurt of social prestige; and since they adopted an admission policy that accepted anyone who applied, they found themselves with a long roll of members, including many with no active interest in natural philosophy. A majority of the early members attended fewer than five meetings and refused to pay the dues, with the result that the Society almost died shortly after its birth. A small minority of men devoted to science, led by Oldenburg and Hooke, kept the society alive. The only purpose in joining was to take part in the meetings, but when the virtuosi were formalizing their meetings into a society, Hartlib was tortured with a medley of diseases that left him practically an invalid. He was therefore physically unable to participate. He died in 1662, less than two years after the society was formed. I feel sure that he would have joined had his health permitted.

"I look upon this society," he wrote of the early meetings of the virtuosi, "as a previous introduction of the grand design here represented." [14] He felt that the realization of the Christian society depended upon a solid foundation of philosophy to which the virtuosi were contributing. Hartlib's letters to Worthington discussing Antilia are at once pathetic and inspiring reading. Already suffering the torments of hell from the medley of ills that soon brought him to the grave, the selfless old reformer revived his spirits with a final thrill of hope. Tottering on the edge of the grave, he was still ready to do battle for his cause. Antilia died of course after the brief rebirth, and Hartlib scarcely outlived it, but the obscure story symbolizes the virtuosi's faith. They lived and worked in the spirit in which Hartlib died, believing that science and religion had no mutual contradictions, believing that the two must advance together toward the victory of truth.

Hartlib's faith in the harmony of science and religion is a constant theme in the works of the virtuosi, always present implicitly, and often expressed explicitly. The virtuosi developed theories about the relations of the two; they demonstrated their concurrence in individual cases; they showed how natural philosophy supports religion. An early virtuoso, John Wilkins, attempted to define the spheres of science and religion and to show their harmony when he came to the defense of Copernican astronomy. An amateur in natural science who made no original discoveries but maintained a lively enthusiasm for science all of his life, and a churchman of deep and enduring faith, Wilkins was well qualified to speak of their relations. While still a young man he published two books arguing the case for Copernican astronomy before the general public—*The Discovery of a World in the Moon* (1638) and *A Discourse Concerning a New Planet* (1640). Only four years before Wilkins began to write, Alexander Ross, the vigilant watchdog of conservatism and orthodoxy, had made a ringing denunciation of the new astronomy in his *Commentum de Terrae Motu;* in the *Discourse Concerning a New Planet* Wilkins vindicated the right of scientific investigation. To the charge made by Ross that the heliocentric hypothesis contradicted passages in the Bible, he replied that natural philosophy cannot be bound

14. Crossley, *The Diary . . . of Dr. John Worthington, 1,* 248–9.

to the biblical text, for the Scripture concerns itself with other matters. While it is supreme and unquestioned as a spiritual authority, the Bible makes no pretense of delivering scientific truths. If it speaks of the sun and the moon rising, it does not intend to comment on astronomy, for the Bible is merely accommodating its language to the vulgar conceptions of its time. No one would dream of maintaining that the value of π is exactly three, he mentioned by way of example, although the Bible reports that the circumference of a certain pool with a diameter of seven feet was twenty-one feet. Obviously there is no attempt to deliver scientific truth in this instance. By following the Bible's exact words, philosophy has admitted a host of absurdities, such as theories about the sphere of upper waters and assertions that the earth is flat. Astronomical truth, Wilkins asserted, is to be found by the study of astronomy; all philosophic truth must proceed from rational inquiry. "There is not any particular by which philosophy has been more endamaged," he declared hotly, "than the ignorant superstition of some men, who in stating the controversies of it, do so closely adhere unto the mere words of Scripture." [15] If both science and religion remain within their own spheres and draw their conclusions from their own valid sources, they can proceed side by side without conflict.

Wilkins did not stop with defending scientific inquiry; he went on to assert that astronomy is a positive aid to religion. Intellectually, its revelation of heavenly order brings us to acknowledge the wisdom of Him who made the heavens. "It proves a God and a providence," he argued, "and incites our hearts to a greater admiration and fear of His omnipotency. 'We may understand by the heavens how much mightier He is That made them; for by the greatness and beauty of the creatures proportionably the Maker of them is seen,' says the book of *Wisdom*. . . . Such a great order and constancy amongst those vast bodies could not at first be made but by a wise providence, nor since preserved without a powerful inhabitant, nor so perpetually governed without a skillful guide." [16] A deeper study of astronomy only reveals the Cre-

15. John Wilkins, *A Discourse Concerning a New Planet* (London, 1640), p. 48.
16. Ibid., pp. 237–9.

ator's excellency the better. Astronomy likewise serves the moral teachings of religion by helping to correct man's estimation of himself and of his actions. When, said Wilkins, rewording the Psalm, I consider the immensity of the universe, and when I consider that I possess a soul of far greater worth than all of this, it must argue a degenerateness and poverty of spirit to busy myself with the narrow and ignoble things that the earth affords. Wilkins maintained, in short, that the study of astronomy and of natural philosophy in general corroborates and strengthens the basic truths of Christianity.

Among Dr. Wilkins' later works only *Mathematical Magic* (1648) was directly related to natural science. Nevertheless his interest in science continued; he met with the Royal Society while he remained in London and took an active part in their deliberations. In a work from the last years of his life, the *Essay towards a Real Character* (1668), he left picturesque testimony that his faith in the harmony of science and religion had not expired since the years of his youth. Using the tables of animals and birds prepared for the *Essay* by John Ray, Wilkins computed the space necessary to house a pair of each with food for forty days—and found that the biblical dimensions of the Ark would suffice with a little to spare.[17] The passage is more quaint than persuasive, to be sure. It even violates his opinion on the proper spheres of science and religion voiced in his books on astronomy, for Wilkins takes up the contention of Alexander Ross that the Bible is scientifically accurate. The very quaintness lights up the conviction that inspired it, however, and within his own mind the two arguments were consistent in expressing the belief that underlay them both, that natural philosophy does not contradict religion.

The same conviction provided the premise for Sir Kenelm Digby's philosophical inquiries, which were published in the *Two Treatises* (1644). Digby intended to demonstrate all of the operations of material bodies and to prove thereby that the soul, which has faculties that cannot be deduced from material operations, must be immortal and immaterial. The conviction inspired John

17. John Wilkins, *An Essay towards a Real Character and a Philosophical Language* (London, 1668), pp. 162–8.

Wallis to his virulent attack on Thomas Hobbes. Wallis set out to lay an axe to the root of atheism by proving Hobbes fallacious on philosophical grounds; and John Ray's belief in the harmony of science and religion led to the publication of his *Physico-theological Discourses* (1692). In order to demonstrate how well the two orders of truth correspond, the three essays that make up the *Discourses* consider biblical revelations against the background of natural philosophy. The first of the essays takes up the Old Testament's story of creation and concludes that natural science does not contradict it. The second is concerned with the deluge and the physical causes that might have produced it. Ray's purpose was not to dispute that the deluge was a miracle—he had no doubt of that—but merely to inquire what natural forces God had employed. The third essay, on the final dissolution of the world, discusses the means by which God might effect the dissolution, whether by fire or flood. In all three of the discourses the matter of fact as found in the Bible is accepted without question, and the arguments from natural philosophy are brought in to show that the authority of the Bible is sound. As in Wilkins' calculations about the ark, no clear distinction separates the position taken by Ray in the *Discourses* from the stand of Alexander Ross in attacking Copernican astronomy. In effect Ray was setting the Bible up as the touchstone of scientific truth. The conviction that natural philosophy and Christianity could not contradict each other had led Ray beyond the limits of his factual knowledge to speculations resting on uncertain foundations. If he stood on shaky ground philosophically, he did demonstrate the firmness of the conviction that had led him there.

On a subject about which he possessed a good deal more information Ray expressed his belief in the harmony of science and religion once more. During his travels on the Continent he became interested in the question of fossils, which posed a number of problems, for religion as well as for science. Ray's effort was to work out an explanation of them that gave offense neither to science nor to religion. That the fossils had been organic life at one time he was sure. Some men asserted that fossils were the products of a plastic power in the earth that molded inorganic matter into forms resembling organic life. Not only did Ray find the answer fallacious scientifically, he considered it dangerous for religion as well. If a plastic power could organize matter into the shapes of living bodies, what was to negate the conclusion that it could create life as well? This solution tended to promote atheism. On the other hand, the theory that fossils had been organic life at one time threatened aspects of the Old Testament. Many fossils exhibited shapes unknown to the living world; they challenged the doctrine, supposedly Christian, of the fixity of species. Fossils of sea life far removed from the sea cast doubt on the commonly accepted age of the world. If land once under the sea had risen to become dry land, it must have been the work of eons; for no physical changes in recorded history have even hinted at such an upheaval. Some earnest defenders of Christianity brought in the deluge to explain the presence of sea shells on dry land. Ray was too honest to accept such a superficial solution. The torrents of the deluge would have washed things down to the sea not up from the sea to the land; and these things would have been scattered everywhere, while the fossils were found in distinct beds, stratified in layers that suggested long periods of deposit. Although he toyed with the problem of fossils for thirty years, Ray never found a satisfactory answer. With our present knowledge of fossils we can say that he was trying to reconcile irreconcilables, but the attempt bore witness to the essential trait of his intellectual make-up. Had he been merely a Christian, he could either have asserted that fossils were nonorganic or have accepted the deluge story without regard for scientific accuracy. Had he been merely a virtuoso, he could have drawn the obvious scientific conclusions, ignoring the contradiction to the Old Testament. Robert Hooke, who was less given to piety, became a major name in early geology by taking the latter course. Since Ray would do neither, he could only leave the question open; but he did nourish a faith that the answer would be found one day, the answer that would at once satisfy the requirements of scientific truth and vindicate the Bible. Expressing his conviction in a letter to Edward Lhwyd, he encouraged the young man in his study of fossils, declaring that his efforts should "in some measure clear up those difficulties wherewith the original of those bodies is entangled, and reconcile all to the novity of the world and Scripture history of the creation and deluge."[18]

On the other hand Ray did not feel obliged to lend the support of science to every superstition that tradition had connected with Christianity. In one sense the whole scientific movement may be looked upon as an attack upon popular superstition, and Ray himself was ruthless in his destruction of old wives' tales. Most of those with which he dealt—stories of strange plants and animals—had no connection with religion, but he did not allow mistaken piety to cloud his opinion of those that did. While he was making his long tour of the Continent, Ray visited the island of Malta, where he found a belief that the absence of serpents proved Malta to be the island on which Saint Paul had been shipwrecked. When a viper had fastened on Paul's hand, according to the story, he had prayed to God that all of the serpents on the island should be turned to stone. Since Malta had no serpents but did have fossils declared to be the remains of serpents, the superstition had grown and flourished. Ray did not hesitate to label it a "Monkish fancy." The fossils were fish bones and could be found in many places where serpents still lived. If the report that there were no serpents on Malta were actually true, he added, "it would be to me a great argument that this was not the island upon which St. Paul was cast when he suffered shipwreck."[19]

In the *History of the Royal Society* (1667) Thomas Sprat took up the theme implied in Ray's story. While maintaining that natural philosophy could not harm true religion, Sprat seized the opportunity to launch a few bolts at enthusiasm and superstition. Natural philosophy, he declared, does not in the least detract from the true prophecies and the miracles attesting to them whereby God has delivered His will to mankind. It is true, however, that science attempts to question all pretended prophecies and miracles in order to prevent superstition from imposing upon

18. Ray to Lhwyd, August 16, 1694; John Ray, *Further Correspondence of John Ray*, ed. R. W. T. Gunther (London, the Ray Society, 1928), p. 252.

19. John Ray, *Observations Topographical, Moral, and Physiological; Made in a Journey through Part of the Low Countries, Germany, Italy, and France* (London, 1673), p. 313.

mankind. Religion is not served by finding a prophet in every madman and a miracle in every unusual event. "It is a dangerous mistake into which many good men fall that we neglect the dominion of God over the world if we do not discover in every turn of human actions many supernatural providences and miraculous events. Whereas it is enough for the honor of His government that He guides the whole creation in its wonted course of causes and effects."[20] Religion is harmed more by polluting it with false witnesses and claims to inspiration and by urging it on the basis of false miracles than by testing all of the claims. Let it be admitted, Sprat granted for the sake of argument, that many virtuosi have been negligent in the worship of God;

> yet perhaps they have been driven on this profaneness by the late extravagant excesses of enthusiasm. The infinite pretences to inspiration and immediate communion with God that have abounded in this age have carried several men of wit so far as to reject the whole matter, who would not have been so exorbitant if the others had kept within more moderate bounds. . . . From hence it may be gathered that the way to reduce a real and sober sense of religion is not by endeavoring to cast a veil of darkness again over the minds of men, but chiefly by allaying the violence of spiritual madness, and that the one extreme will decrease proportionably to the lessening of the other.[21]

Sprat's purpose was to cast Anglican ridicule on the Puritans, but his words may be given a broader meaning, to which most of the virtuosi would have assented. When they argued that natural philosophy cannot contradict Christianity, the virtuosi had in mind basic doctrines of Christianity that they considered essential. Different men might consider different things essential, but all were agreed that the conclusions of science did not invalidate religion. Not everything could escape their criticism by masquerading as Christian doctrine, and in effect the virtuosi demanded the right to define the essentials for themselves. They regarded

20. Thomas Sprat, *The History of the Royal Society of London for the Improving of Natural Knowledge* (London, 1667), p. 360.

21. Ibid., pp. 367–8.

their studies as an aid to religion, enriching it by purifying it and strengthening it by cutting away parasitical growths.

Sprat maintained, as Wilkins had done before him, that the study of nature helps Christianity by confirming and strengthening the principles of natural religion. The wonders of creation reveal the power and wisdom of God and encourage us to worship Him. Hence the virtuoso will be led "to admire the wonderful contrivance of the Creation, and so to apply and direct his praises aright, which no doubt, when they are offered up to heaven from the mouth of one who has well studied what he commends, will be more suitable to the Divine Nature than the blind applauses of the ignorant."[22] Meanwhile the study of nature does nothing to destroy the theological doctrines that derive from the Scriptures. These doctrines are either comprehensible by reason, and therefore unable to be overthrown by natural philosophy, or above reason, and so dependent upon the miracles that declare the Bible to be the word of God. Because of natural philosophy's knowledge of phenomena, it can best demonstrate the validity of the miracles on which revelation depends. Sprat's opinion was echoed by other virtuosi. Indeed Walter Charleton found it so acceptable that he repeated the exact words in his *Three Anatomic Lectures* (1683).[23] The fullest development of the argument was found in the writings of Robert Boyle, who did more than any other virtuoso to justify the religious value of scientific investigation.

Boyle is the foremost example that the 17th century can offer of scientific investigation impinging on the Christian consciousness. History has known Boyle primarily as a scientist, the father of chemistry and the discoverer of Boyle's Law. In the eyes of his contemporaries he was no less notable as a Christian gentleman. Religion rather than science was the foundation of his being. At an early age Boyle underwent a "conversion," a terrifying experience of contingent existence hanging over the abyss. Possibly the sudden conversion, as told some years later in an autobiographical sketch, was only the most vivid event in a protracted chain of experiences. In any event, before he was well out of ado-

22. Ibid., p. 349.
23. (London, 1683), pp. 37–8.

lescence, Robert Boyle had centered his whole life on Christian practice. Governing his conduct by the strictest puritanical code, he abstained from tobacco, alcohol, excesses in any form. It is reported that he never uttered the name of God without first pausing reverently. At times Boyle's piety verged on mere primness. It is more painful than inspiring to find him, as a young man under twenty, urging his older brother to reform or composing a pretentious treatise against profanity; and it would be reassuring to hear him utter one good mouth-filling oath. Nevertheless a genuine nobility redeemed him from priggishness. If Boyle was prim and overly nice, he was at least not a hypocrite. In honesty and righteousness he governed his life by the code he preached, the religious germ leavening his every act. Natural philosophy, the second passion of his life, was so closely bound up with Christianity that the two coalesced in his mind. As Bishop Burnet said of his scientific researches, "his main design in that, (on which, as he had put his own eye most constantly, so he took care to put others often in mind of it) was to raise in himself and others vaster thoughts of the greatness and glory and of the wisdom and goodness of God." [24] Robert Boyle may be named with his own book the Christian Virtuoso.

More perhaps than any other virtuoso Boyle sought, and found, the hand of God in the creation. Through all of his works runs a never silent melody of enraptured surprise at the Creator's ingenuity. The more deeply he probed into nature, the more humbly he acknowledged her Maker. Here was a piece of workmanship without flaw, rich in its intricate detail beyond the imagination of man. Truly nature proclaims the glory of God. Boyle never lost the sense of wonder that his first apprehension of natural glory awoke. Throughout his life, from his adolescent exercises in piety to his last philosophical book, he sang hymns to the glorious Creator. In line after line, paragraph after paragraph, page after endless page, he developed the theme, systematized and enriched it, worked in variations, and returned to the original to sound it and sound it again until the reader would gladly admit the point if only to escape further punishment. The creation be-

24. Gilbert Burnet, *Lives, Characters, and an Address to Posterity,* ed. John Jebb (London, 1833), p. 345.

speaks an intelligent and powerful Creator. His imprint is heavy upon it. Boyle reaped an unfailing harvest of piety from his scientific investigations.

God displayed His power to Boyle in the size and vastness of the universe. Boyle noted the earth—it contains 10,882,080,000 cubic miles of solid material, yet is but a speck in the immensity of space. He considered the quantity of motion in the rotation of the earth, the movements of the heavenly bodies, and the agitation of invisible corpuscles. He counted the number of creatures inhabiting the earth, in variety extending from the elephant and whale down to microscopical creatures, all created from nothing by God. Such was the power of God in his eyes.[25] The Almighty revealed His wisdom in the ordered harmony of the multitude of creatures. Despite their diversity, all fit together in a cooperative union, each helping the others and contributing to the perfection of the whole. Every creature displays the same organic unity in microcosm, each part delicately formed yet fitted harmoniously to the others. The human body especially stimulated Boyle's imagination, and among its parts the eye attracted his attention most strongly. Could man conceive of such a mechanism, much less construct it? Boyle was forced to answer that man could not. The riches of nature exceed the imagination of man.[26] The eye, the body, the variety of creatures, the scope of the heavens, all nature displays the power and wisdom of God if man will but trouble to look. "For the works of God are not like the tricks of jugglers, or the pageants that entertain princes, where concealment is requisite to wonder; but the knowledge of the works of God proportions our admiration of them, they participating and disclosing so much of the unexhausted perfections of their Author, that the further we contemplate them, the more footsteps and impressions we discover of the perfections of their Creator; and our utmost can but give us a just veneration of His omniscience." [27] Robert Boyle's attitude toward nature never lost the element of surprise which has already been noted in other virtuosi. To him the natural

25. Boyle, *Works*, 2, 10–14, 20–5; 5, 132–5.
26. Ibid., 2, 44–63; 5, 135–9.
27. Ibid., 2, 30.

order remained a treasure of endless curiosities and wonders on which his pious soul could gaze forever.

Nevertheless, Boyle reported, some divines urged men to reject the study of nature lest they promote disbelief. He set out to prove how wrong they were. The proposition was absurd: as well promote atheism by the study of the Bible. Developing a theme from Bacon, Boyle asserted that nature is God's revelation as much as the Bible; His message can be found in either place. Preferably His message can be found in both places, for in neglecting natural philosophy men shun an essential religious duty. In the *Usefulness of Experimental Philosophy* (1663), the first work on natural philosophy that he wrote—though it was not the first that he published—Boyle developed his argument. Since the Almighty created the universe to display His glory, He did not intend His workmanship to be ignored. While beasts inhabit and enjoy the world, man is required to study it. "It is the first act in religion and equally obliging in all religions; it is the duty of man as man, and the homage we pay for the privilege of reason, which was given us, not only to refer ourselves, but the other creatures that want it, to the Creator's glory." [28] The vulgar astonishment of an unlettered boor, whose wonder may stem from mere ignorance, glorifies God less than the learned hymns of men who fully understand what it is they praise. In the homely phrase which Boyle used to sum up his case, "on the opened body of the same animal a skillful anatomist will make reflections as much more to the honor of its Creator than an ordinary butcher can, as the music made on a lute by a rare lutanist will be preferable to the noise made on the same instrument by a stranger unto melody." [29] The study of nature is not only permitted, it is demanded as a religious duty. Man neglects it at his peril.

Boyle took up the question once more in his last important work, *The Christian Virtuoso* (1690), written, as he said in the preface, to show that there is no inconsistency between Christianity and scientific investigation. The virtuoso, he argued, becomes a better judge of religious truth through his study of philosophy.

28. Ibid., pp. 62–3.
29. Ibid., pp. 63.

Accustomed to seeking truth and to investigating abstruse problems, he is less apt to be swayed by arguments against Christianity because he has seen newly discovered facts upset old arguments against the mechanical philosophy. Of course Boyle's reasoning would apply equally well to arguments in favor of Christianity, but it did not occur to him to question the truth of Christianity. Essentially the advocate who did not search for truth with an unbiased mind, he endeavored only to uncover evidence supporting Christianity. The virtuoso, he said, will check the proofs of Christianity more thoroughly than the ordinary man; since Christianity is true, the virtuoso's acceptance will be more sure than the ordinary man's. In pursuit of his own special interest the virtuoso will penetrate below the surface of natural phenomena to discover "more curious and excellent tokens and effects of divine artifice in the hidden and innermost recesses of them [which] are not to be discovered by perfunctory looks of oscitant or unskillful beholders, but require, as well as deserve, the most attentive and prying inspection of inquisitive and well-instructed considerers."[30] In his argument Boyle did much to reveal the mental processes of the virtuosi. While he set up theories about the relationship of science and religion, in the end they rested not on their formal reasoning but on Boyle's intuition of natural glory, on his perception of "curious and excellent tokens and effects of divine artifice." The farther he penetrated into nature, the more splendid were the things that he discovered. Atheism or skepticism were impossible to his mind, for each fresh observation awoke more profound reverence for the Creator. Coming to Mother Nature in search of spiritual food he was bountifully supplied; and he concluded that he had found the basic diet that sustains all religion.

Boyle was not alone in raising the study of nature to a positive religious duty. The greatest naturalist of the age, John Ray, fused his love of nature with religious zeal and, following the lines of Boyle's development, erected his personal response to nature into a general theory of devotion. In the preface to his first book, *Catalogus Plantarum circa Cantabrigiam nascentium* (1660) Ray told how he became absorbed in natural history. Whenever he

30. Ibid., 5, 516.

rested from his studies or walked in the country, he was compelled to admire the variegated beauty of nature. The charms of spring enthralled him. The more he gazed, the more his wonder grew; and he was drawn to the study of plants as Ovid was drawn to verse. In "the vast library of creation," as he called nature, he found an unlimited store of divinity. For a free man, he declared, there is "no occupation more worthy and delightful than to contemplate the beauteous works of nature and honor the infinite wisdom and goodness of God." [31] Benjamin Allen reported Ray to have said that "a spoil or smile of grass showed a Deity as much as anything." [32]

In the *Wisdom of God Manifested in the Works of the Creation* (1691), an immensely popular book which ran through four editions while the author was still alive in the decade following its first publication and through many more in the years after his death, the pious naturalist poured forth his song of praise. A passage from Psalm 104 introduced the argument—"How manifold are thy works, O Lord! In wisdom hast thou made them all"—and in the pages following, the author documented the text. The number of the Lord's works, beyond the power of man to investigate, demonstrates the extent of the Creator's skill and the wealth of His power and wisdom. In the heavens the stars are innumerable; each improvement in the telescope reveals multitudes more. Every star, it would now appear, has planets circling about it, and every planet its complement of corporeal creatures in as great a variety as the earth. On the earth alone dwell at least a hundred and fifty species of beasts, some five hundred species of birds, perhaps three thousand fish, twenty thousand insects, and more plants. If the number of creatures be so exceeding great, how immense must be the power and wisdom of Him Who made them? "For . . . as it argues and manifests more skill by far in an artificer to be able to frame both clocks and watches, and pumps and mills, and granadoes and rockets, than he could display in but making one of those sorts of engines, so the Almighty discovers more of His wisdom in forming such a vast multitude of different sorts

31. Cited in Raven, *Ray*, p. 83.
32. Ibid., p. 9.

of creatures, and all with admirable and irreprovable art, than if He had created but a few; for this declares the greatness and unbounded capacity of His understanding." [33]

A lifetime of study did not succeed in making Ray indifferent to the wonders of nature. Always he found fresh treasures to stir his mind and soul. "You ask what is the use of butterflies?" he inquired. "I reply to adorn the world and delight the eyes of men, to brighten the countryside like so many golden jewels. To contemplate their exquisite beauty and variety is to experience the truest pleasure. To gaze inquiringly at such elegance of color and form devised by the ingenuity of nature and painted by her artist's pencil, is to acknowledge and adore the imprint of the art of God." [34] A spoil or smile of grass did indeed show a God.

With such a response to a butterfly Ray's natural philosophy was scarcely distinguished from religion. In the *Wisdom of God* he argued that because the Almighty created man able to study nature, He intended that man ought to study nature. The pursuit of natural philosophy is a religious duty. Since no creature on the earth besides man is capable of admiring the divine handiwork, man frustrates one end of creation if he ignores it. Those who shirk the task, he stated, "do as it were rob God of some of His glory, in neglecting or slighting so eminent a subject of it, and wherein they might have discovered so much art, wisdom, and contrivance." [35] With Boyle John Ray elevated the investigation of nature to the level of a fundamental religious act, preparing the soul for worship in the formal Christian church.

Beyond the works of Boyle and Ray, the defense of science as a religious study received its fullest expression in the essays of Joseph Glanvill. Although Glanvill was not himself the author of any scientific work of significance, he did act as spokesman for the virtuosi's attitude. When he praised the virtuosi for demonstrating the harmony of science and religion, he denounced those who maintained that the study of nature must lead men to give up religion. The latter opinion, implying that religion is built on

33. John Ray, *The Wisdom of God Manifested in the Works of the Creation* (4th ed. London, 1704), p. 26.

34. Cited in Raven, *Ray*, p. 407.

35. Ray, *Wisdom of God*, p. 210.

a foundation of ignorance, could only result in the triumph of atheism.[36] In other essays Glanvill asserted the religious value of natural philosophy. Man, he declared, is commanded to praise God, and those who would praise Him must study His works.

> His works receive but little glory from the rude wonder of the ignorant, and there is no wise man that values the applauses of a blind admiration. No one can give God the glory of His providence that lets the particulars of it pass by him unobserved, nor can he render due acknowledgments to His Word that does not search the Scriptures. 'Tis equally impossible to praise the Almighty as we ought for His works while we carelessly consider them. We are commanded to search for wisdom as for hidden treasure; it lies not exposed in the common ways, and the chief wonders of Divine art and goodness are not on the surface of things laid open to every careless eye. The tribute of praise that we owe our Maker is not a formal slight confession that His works are wonderful and glorious, but such an acknowledgment as proceeds from deep observation and acquaintance with them. And though our profoundest study and inquiries cannot unfold all the mysteries of nature, yet do they still discover new motives to devout admiration and new objects for our loudest praises.[37]

"The wonders of the Almighty are not seen but by those that go down into the deep," he remarked in another passage, summarizing both the piety with which the virtuoso approached his work, and the religious response which it inspired in him.[38]

Boyle, Ray, and Glanvill spoke for the virtuosi as a whole when they found in natural philosophy a basic religious duty, for the virtuosi considered the study of nature to be a spiritual exercise and a religious experience. From the wonders of nature they drew inspiration which confirmed religion and accorded with the teach-

36. Glanvill, "An Address to the Royal Society" in *Scepsis Scientifica.*

37. Joseph Glanvill, *Essays on Several Important Subjects in Philosophy and Religion* (London, 1676), pp. 5–6.

38. Joseph Glanvill, *The Vanity of Dogmatizing: or Confidence in Opinions* (London, 1661; reproduced by Columbia University Press for the Facsimile Text Society, 1931), p. 246.

ings of the Christian church. In the 18th century Hume was to show that the argument from the creation to the Creator was unsound; but whether or not their reasoning was logically valid, the heavens did declare the glory of God to the virtuosi. Profoundly moved, as they were, by the works of nature, they could not believe that their studies were irreligious. The hounds of superstition might howl; the virtuosi proceeded unabashed. Asserting that the revelation in nature could not contradict the written Word, they not only defended science but praised it as a religious act. All truth is one, they were saying; natural philosophy does not and cannot contradict Christianity.

CHAPTER 3

The Harmony of Existence

> I shall remark the care that is taken for the preservation of the weak . . . for as it is a demonstration of the divine power and magnificence to create such variety of animals . . . so is it no less argument of His wisdom to give to these means, and the power and skill of using them, to preserve themselves . . .
>
> . . . those vastly small animalcula, not to be seen without a microscope, with which the waters are replete, serve for food to some others of the small insects of the waters . . . This to me seems a wonderful work of God, to provide for the minutest creatures of the waters food proper for them, that is, minute and tender, and fit for their organs of swallowing.
>
> Ray, *The Wisdom of God Manifested in the Works of the* Creation

THE VIRTUOSO's disposition to react emotionally to nature and to see it as God's direct work was conditioned by the fact that he was already a Christian. Convinced from the beginning that the Almighty's fingers had molded creation, he studied nature determined to admire the divine handicraft. More than a demonstration that nature reveals the Creator, his work is proof that anyone bent on the worship of God can force the creation to reveal Him. The Christian virtuoso had been a Christian before he became a virtuoso. Born into a Christian society, reared in an atmosphere heavy with Christian teachings, brought to maturity, in many instances, during the great Civil War when religious passions drove men to risk life and position for their faiths, he inevitably had the lessons of Christianity driven deeply into his soul; and Christian beliefs played a major role in shaping the outlook that he brought to the study of nature. By coloring his attitude toward nature his religious preconceptions subtly influenced his idea of nature. The virtuoso's conviction that nature was governed by law was less the conclusion that he reached through science than the premise that he brought from Christianity.

His Christian training led him to see wise planning where possibly chance had ruled. It led him to impose a pattern of fatherly benevolence on nature which excluded from view all evidence of squalor and suffering. Given the virtuoso's rosy conception of natural harmony, the conclusion that divine beneficence created the world followed inevitably. Blind matter moved by impersonal laws could not have produced an order such as he pictured. Unconsciously the virtuoso was reasoning in a circle, for his original conception of the natural order derived from his belief in the Creator's goodness. Religion influenced the conclusions of science in its early days as much as science influenced religion. The Christian doctrines of the creation and of the Creator's omniscience were primary sources of the virtuoso's idea of nature as an ordered whole. The very zeal that he displayed in forcing every event into the pattern of harmony shows how religious beliefs could lead him beyond the conclusions of scientific investigation alone.

The most common argument used by the virtuoso to prove the existence and wisdom of God cited the evidence of His skill in designing His creatures. Every element of nature, the virtuoso believed, from the heavenly spheres down to the parts of insects, fits exactly into the notch it fills in the universal order. The virtuoso might argue, for instance, that the human leg is formed perfectly for walking. Without joints at the knee and the ankle a man's movement would be stiff and slow. Were the muscles less strong, they could not carry the body; were they larger, the added strength would be wasted. In a word, the virtuoso found teleological significance in everything. Where the age following Darwin has come to think of evolution, the 17th century for the most part, despite the admonition of Descartes, thought of original design. If the leg is fit for walking, the Creator must have made it for walking. An anecdote from Pepys' diary illustrates the virtuoso's orientation. Pepys tells us of Walter Charleton's argument that nature fashions every creature's teeth to correspond to the food she intends for them. Lord Brouncker objected that the process works in reverse; creatures find the food proper for their teeth. Charleton rejoined that even before they learn by experience, animals naturally prefer one type of food to another.[1] Brouncker was upholding a position that assumed the adaptation of creatures to their environment, while Charleton was working on the assumption of an original creation by design. One step more would have carried him on to admire the wisdom of God in making each of the countless varieties of teeth just right for the food it must chew, and the manifest adaptation of teeth to food would have proved to him that the world was created by an omniscient God. Every virtuoso who expounded natural religion took this step. *He proved the existence of God from assumptions in which the existence of God was already implicit.* It is pointless to condemn the virtuoso for not understanding evolution; at the same time it is clear that his premises already held his conclusions.

Despite the virtuosi's sustained and vigorous denunciation of Scholastic philosophy, they drew heavily upon the medieval heritage in their use of teleology. Spurning the rigorous implications of the mechanical conception of nature which Descartes and Spinoza drew in rejecting the investigation of final causes, the English virtuosi, almost without exception, refused to look upon nature as an impersonal machine. Inconsistent though they may have been, they combined the mechanical view of nature with the medieval conception that nature is the product of divine goodness. A deep-seated conviction that creation is a benevolent order reflecting divine goodness meant that in their minds the seemingly inexorable cosmic machine lost its harsh and inhuman aspect because it ran with the lubricant of infinite love. Thus the virtuosi's conception of nature was an amalgam of the new mechanical hypothesis and the Christian philosophy of the Middle Ages, an amalgam which made it possible for them to employ the arguments of teleology without conscious inconsistency. There was a fundamental difference which separated the virtuosi's usual employment of teleology from the medieval use of it. Regardless of their eye for the purposes of things, the virtuosi's ultimate question was not "why" but "how." How does the universal machine operate? How does this individual part contribute to the operation of the whole? The strong utilitarian motive that helped to

1. Cited in editor's preface; Walter Charleton, *Epicurus's Morals: Collected and Faithfully Englished,* ed. Frederic Manning (London, 1926), p. ix.

foster the growth of modern science directed attention away from ultimate explanations and toward descriptive knowledge of natural phenomena, which would give power to control nature. In England the legacy of Sir Francis Bacon made the utilitarian motive particularly strong. Even in questions so far removed from worldly utility as proofs of God's existence, the virtuosi used teleological arguments that demonstrated how individual parts contributed to the operation of a perfect machine. Just as they saw God in the role of original first cause rather than immanent final cause, so they saw nature as a complete and functioning machine which is not striving toward a final end. Nature is not tending toward perfection and goodness; it is perfect and good. To apply Boyle's omnipresent analogy, the virtuosi looked for the purpose of each wheel in the finished—and benevolent—clock. Actually the virtuosi implied an answer to the ultimate "why" of creation. The mechanical simile, to which they invariably reverted when searching for an explanatory analogy, assumed that the creation of a perfect machine was the proper means for God to demonstrate His wisdom. The smooth operation of the cosmic clock was a final end in itself, and the goal of the virtuoso was to describe how it worked—and to demonstrate the infinite skill of the Divine Clockmaker at the same time. One exception, in which their use of teleology most closely approximated medieval philosophy, should be noted; it is also an inconsistency (dictated largely by Christian influences) with the mechanical conception of nature: the Christian conviction that man is the special darling of creation, together with the normal homocentric tendency of human thought, made it possible for the virtuosi to think that the major purpose of the rest of nature was the convenience of man. Nature is a machine, but the machine is meant to serve its most important part, the human species. It was particularly easy for the virtuosi to accept this element of medieval teleology because in this argument Scholastic philosophy most nearly approximated the utilitarianism of 17th century science. "The world was made for man, Hunt, not man for the world," Bacon had told his servant. The drive to control nature could readily accommodate itself to the idea that the rest of creation is subordinate to human utility.

Robert Boyle devoted countless pages to admiring God's wisdom in the creation. Scarcely a work of his did not pause at some point to demonstrate how nature reveals her maker. In one essay, *A Disquisition about Final Causes* (1688), he tried to analyze the limits within which teleological arguments can safely be used. Descartes had declared it to be presumptuous for man to investigate God's final causes. If he meant that man cannot discover all of God's ends, Boyle agreed with him; but if Descartes meant that man cannot know any of God's purposes, Boyle objected strongly. Some final causes are so obvious—the eye, for instance, is clearly designed for sight—that to miss them would be folly and blindness. Boyle believed that the Christian philosopher especially must observe final causes, because they are a powerful argument for the existence of God.[2] On the other hand he recognized that teleology could be—indeed had been—carried to the bounds of absurdity, and he surveyed nature to mark out those areas where its use is safe. The heavenly bodies are of some service to man, but it is doubtful whether they do not fulfill a higher end as well. One must beware of asserting their final cause. The ends of most inanimate objects are obscure at best; while it is not unreasonable to think that they were made for particular purposes, they may well have been the chance productions of matter in motion. With living creatures final causes become manifest. Mere chance could not have produced such intricate machines with parts nicely proportioned to the whole and to the particular circumstances in which the creature must live.[3] While he added an injunction that we should not be too hasty and positive in assigning final causes, Boyle himself launched into a rhapsody of praise for the Creator's skill. In both this essay and others Boyle gave his highest admiration to the eye. Each eye in itself is a delicate mechanism containing every part needed for sight, and in every seeing creature it is modified to fill a special necessity. With the assumption that sight was decided upon first, Boyle praised the Creator's success in designing a machine for the job. Other virtuosi were to repeat his reasoning; few of them observed his caution about inanimate bodies.

In the mutual adjustment of creatures to form a harmonious

2. Boyle, *Works*, 5, 397–401.
3. Ibid., 5, 420–39.

universe, as well as in the contrivance of their individual parts, Boyle found evidence of the Creator's goodness. God, he declared, has qualified His works "to be wonderfully serviceable to one another, and a great number of them to be particularly subservient to the necessities and utilities of man." [4] This happy bit of optimism tinted and softened Boyle's mechanical conception of nature. When he was remarking once on the harmony within the creation, he called attention to the good fortune of young lambs to be born in the spring when there is fresh young grass for them to feed on. Similarly, silkworms are hatched when the tender young mulberry leaves come out.[5] Here indeed was a picture of nature in which the lion and the lamb lay down together. All of creation was the embodiment of divine love, and every creature dwelt in a bliss of neighborly cooperation. Should the lion have eaten the lamb, the same lamb so providently supplied with fresh new grass, Boyle would have seen only the goodness of God in providing the lion with food. The examples of the lamb and the silkworm are not wholly unlike the evidence that Darwin was to use, but how greatly the two men differed! Where the champion of evolution found nature red in tooth and claw, lamb vying with lamb for the blade of grass, grass vying with grass to survive the depredations of the lamb, the Christian virtuoso saw only love and harmony and the providence of God.

The eminent doctor Walter Charleton, the virtuoso whom Pepys reported as arguing that teeth were designed for the food they would chew, displayed the same sturdy determination to locate divine purpose in every aspect of nature. In an attack on materialism, *The Darkness of Atheism* (1652), he applauded the wisdom that God had displayed in the location of sun and earth. If the sun were any closer, he declared, we should all burn up, and a year might be shortened to scarcely a month. If, on the other hand, the Creator had placed the sun at a greater distance, we should be too cold, and a year would be too long. The omniscience of God manifested itself in making the distance just right.[6] Where

4. Ibid., p. 398.
5. Ibid., p. 138.
6. Walter Charleton, *The Darkness of Atheism Dispelled by the Light of Nature. A Physico-theological Treatise* (London, 1652), pp. 57–60.

the age following Darwin finds adaptation of life to existing conditions, Charleton could see only creation of conditions suitable to life as it is.

Charleton's *Darkness of Atheism* contains a discussion of divine providence pregnant with and founded upon the conviction that nature is an ordered cosmos. His case for providence began with a discussion of the nature of God. Since He is omnipotent, God is well able to regulate and rule all that he has created. Since He is supreme in goodness, He will preserve His creation from the chaos and destruction that must inevitably ensue if it be left to run by itself.[7] Both arguments rest on the premise that the universe is in fact an ordered and coherent whole, worthy of the God Who maintains it. Drawing then upon natural philosophy, Charleton sought to demonstrate that nature does reveal such a pattern. God is both able and willing to rule His creation, Charleton was saying; if we examine nature we shall see that in fact He does rule it.

> For whoever (though a mere pagan, whose brain never received the impression of either of those two notions, "Creator" and "Providence") shall speculate the world in an engyscope or magnifying glass, i.e., shall look upon it in the distinction of its several orders of natures, observe the commodious disposition of parts so vast in quantity, so infinite in diversity, so symmetrical in proportion, so exquisite in pulchritude; shall contemplate the comeliness, splendor, constancy, conversions, revolutions, vicissitudes, and harmony of celestial bodies; shall thence descend to sublunary, and with sober admiration consider the necessary difference of seasons, the certain-uncertain succession of contrary tempests, the inexhaustible treasury of jewels, metals, and other wealthy minerals concreted in the fertile womb of the earth, the numerous, useful, and elegant stock of vegetables, the swarms of various animals, and in each of these the multitude, symmetry, connection, and destination of organs; I say, whoever shall with attentive thoughts perpend the excellencies of the inimitable artifices (for all things are artificial, nature being

7. Ibid., pp. 107–10.

> the art of God) cannot, unless he contradict the testimony of his own conscience and invalidate the evidence of that authentic criterion, the light of nature, but be satisfied that as nothing less than an infinite power and wisdom could contrive and finish, so nothing less than the incessant vigilancy and moderation of an infinite providence can conserve and regulate them in order to the mutual benefit of each other, and all conspiring, though in their contentions, to the promotion of the common interest.[8]

Not only delight in nature's beauties and wonder at her richness shines through Charleton's statement, but also unquestioned faith in her benevolent order. It is not a reasoned conclusion but a basic attitude, a presupposition, a deposit from centuries of Christian thought.

Charleton was aware, of course, that all was not sweetness and harmony. Anomalies, monsters, irregularities blot the face of nature. Can we praise an order disordered by such events? "Does not irregularity render order the more conspicuous and amiable," Charleton asked in response, "and deformity, like the negro drawn at Cleopatra's elbow, serve as a foil to set off beauty? Are not the moles on the cheeks of nature, as those on Venus's skin, placed there to illustrate or whiten the snow and sweeten the feature of her face? . . . Does not the painter then show the most of skill when he refracts the glaring luster of his lighter colors with a veil of sables, and makes the beauty of his piece more visible by clouding it with a becoming shadow?"[9] Every occurrence in the world is the product of God's mature judgment, he added in a phrase that changed the metaphor; "though in the ears of man they may sound discords to the music of particular natures, yet will they at last be found well composed airs necessary both to sweeten and fill up the common harmony of the universe."[10] The argument then took two forms: the world is good and therefore is ruled by God; the world is ruled by God and therefore is good. On the one hand there are beauty and harmony which reveal

8. Ibid., p. 115.
9. Ibid., p. 127.
10. Ibid., p. 126.

the benevolent Father; on the other, there are shadows and discords which do not lead us to doubt the Father but only teach us that His wisdom exceeds ours.

However much Charleton tried to prove that divine providence rules the world, in the end he could only assert it. So immense are the bounds of the world, he suggested in one passage, so numerous its divisions, and each of the divisions peopled with so many millions of different and discordant natures, "that no reason can admit it so much as probable that a constant correspondence could be maintained, and a general amity observed through all, without the conserving influence of a Rector General or Supervisor, Whose will receives laws from His wisdom and gives them to all besides Himself."[11] Charleton wrote these words in 1652 when the flower of modern science was yet a bud. Copernicus and Kepler had propounded basic theories in astronomy. Galileo had done the same for mechanics. Harvey's discovery of the blood's circulation had opened a new era in physiology. Nevertheless, much work in all of these fields and many discoveries still unsuspected were necessary to bring science to the full bloom which the later 17th century was to witness. Charleton's statement of nature's consistency and law appears as a declaration of faith more than a scientific conclusion, a declaration of faith which could inspire the search for laws both by Charleton and by other virtuosi.

Nearly every book published by the virtuosi revealed a similar faith in the order of nature. It appeared in a variety of forms. It might be used, as it was in the passage quoted from Charleton, to explain away unpleasant facts. If anomalies seemed to upset the natural order, they could be blamed on the limitations of the human intellect. Faith in the natural order suggested that a higher intelligence, able to comprehend the whole universe, would see in apparent anomalies perfectly natural events, which were strange only to human, and therefore restricted, comprehension. Used in this form it was closely associated with the concept of the great chain of being. On the other hand, belief in the order of nature could become a premise for conclusions in natural philosophy. John Wilkins, for instance, made it the basic argument of his

11. Ibid., p. 113.

famous book *The Discovery of a World in the Moon* (1638). He reasoned that the moon is solid, opaque, and nonluminous, properties which it has in common with the earth. There are mountains on the moon as telescopes reveal, and the spots are apparently seas. It has an atmosphere and a succession of seasons. In all important matters, then, the moon is similar to the earth and is therefore habitable. Nature would produce nothing without an end; and the purpose of the moon, as has been shown, is to be inhabited. Therefore the moon is a world—a sphere on which men live.[12] In his second treatise on astronomy, *A Discourse Concerning a New Planet* (1640), Wilkins stated his rationale explicitly. While the *Discovery of a World in the Moon* maintained that the earth is not unique, that at least one other body is inhabited, the *Discourse Concerning a New Planet* openly supported the new astronomy by demonstrating the revolution of the earth around the sun. Following a common argument for the Copernican system, Wilkins suggested the improbability of the heavens moving at the tremendous speed necessary if they were circling the earth each day. Of course, God has the power to move the heavens that fast. "But, however," Wilkins concluded, reaching the foundation of his case, "the question here is not what can be done, but what is most likely to be done according to the usual course of nature. 'Tis the part of a philosopher, in the resolution of natural events, not to fly unto the absolute power of God and tell us what He can do, but what according to the usual way of providence is most likely to be done, to find out such causes of things as may seem most easy and probable to our reason." [13]

The source of the virtuosi's faith in the natural order reveals itself in their religious writings, where a similar premise supported their reasoning. Parallel to John Wilkins' demonstration that the moon is a world runs his argument for the immortality of the soul. He began with the proposition that God has created an object to satisfy every affection in nature. If the natural human desire for eternal life has no fulfillment, man is duped and nature is im-

12. John Wilkins, *The Discovery of a New World; or, a Discourse Tending to Prove That 'Tis Probable there May Be Another Habitable World in the Moon* (3d impression, London, 1640), passim. The first edition in 1638 bore the better known title *The Discovery of a World in the Moon.*

13. Wilkins, *Discourse Concerning a New Planet,* p. 193.

perfect. Therefore the soul is immortal. Again, the case was stated on moral grounds, showing that the reward of virtue and the punishment of vice require a future life. In both arguments Wilkins assumed a rational and moral order in the spiritual realm that foreshadowed the order of the physical world. His reasoning followed a standard demonstration for eternal life which was repeated by every virtuoso who considered the subject. Such diverse figures as Kenelm Digby, Walter Charleton, and Thomas Sydenham, for instance, all employed it without questioning its premise.

The assumption underlay much of their thought about spiritual matters. Joseph Glanvill, for instance, stated it explicitly in discussing the possibility of divine care and protection of men. "Providence," he confessed, "is an unfathomable depth; and if we should not believe the phenomena of our senses before we can reconcile them to our notions of providence, we must be grosser skeptics than ever yet were extant." The miseries of the present life, the unequal distribution of good and evil, the ignorance and barbarity of the greatest part of mankind, the fatal disadvantages we are all under, and the hazard we run of being eternally miserable and undone are things that can hardly be made consistent with that wisdom and goodness that we are sure has made and mingled itself with all things. "And yet," he continued, "we believe there is a beauty and harmony and goodness in that providence, though we cannot unriddle it in particular instances, nor, by reason of our ignorance and imperfection, clear it from contradicting appearances." [14] Here surely is the prototype of the virtuoso's faith in the natural order. As the Christian tradition taught him that God would not create in man the desire for eternal life without satisfying it, taught him that justice will not forever go unvindicated, taught him that divine providence is a reality whatever the slings of outrageous fortune, so it taught him that below the seeming diversity of nature lies an ordered pattern created by God.

To disclose the pattern was the object of every virtuoso who set out to prove that nature reveals her Creator. Some limited themselves to asides dropped into scientific treatises. Richard

14. Joseph Glanvill, *A Blow at Modern Sadducism in Some Philosophical Considerations about Witchcraft* (London, 1668), p. 25.

Lower's *De Corde* (1669), for example, found evidence of divine planning throughout the circulatory system. He pointed out that the left ventricle of the heart, which must drive the blood through the whole body, is much more muscular than the right ventricle, which only forces it through the lungs. Valves in the veins permit flow in one direction, but prohibit it in the other. Even a thing apparently so arbitrary as the network of arteries and veins turns out, upon inspection, to be the product of wisdom and design. At the exit from the left ventricle the blood is thrown straight up; if nothing interposed to change its direction, too much would go to the brain and too little to the body. The "divine Artificer" foresaw the problem and provided for it in the construction of the aorta. In animals, whose hearts move rather forcibly, the blood does not run straight into the axillary and cervical arteries but first turns through part of a circle in the aortic arch, directing the main flow of blood downward.[15] Lower injected the spirit that informs this example into the entire book, finding every part of the heart not merely workable but perfect, the product of supreme intelligence creating a cosmos.

While the theme remained an undercurrent in Lower's work, it swelled to the surface and dominated the argument in two books that undertook to display systematically the divine pattern of nature. The first part of Nehemiah Grew's *Cosmologia Sacra* (1701) devoted itself to this task. After noting the excellent organization of the universe, Grew concentrated upon the uses of its parts. Every member of nature has its use, he maintained, and all are exactly fitted for the use they serve. "The water flows, the wind blows, the rain falls, the sun shines, heaven and earth act and move, and all plants live and grow for the use and benefit of sensible creatures, and all inferior creatures for the service of those above them. Nor is there any one of so many parts which compose every creature but what is either necessary for its being or convenient for its better being. As it has nothing hurtful or redundant, so no agreeable part is wanting to it." [16] To substantiate

15. Richard Lower, *Tractatus de Corde,* trans. K. J. Franklin, in Gunther, *Early Science in Oxford,* 9, [52].

16. Nehemiah Grew, *Cosmologia Sacra: or a Discourse of the Universe as It Is the Creature and Kingdom of God* (London, 1701), p. 23.

his statement Grew called upon his own extensive knowledge of organic life. The manifold variety of ears illustrate how the Creator has adapted each animal's structure to the position it must fill. Since the owl sits on branches and listens for things below, its ear projection is above the auditory canal; while the fox which scouts for roosting prey has ears which are directed upward. The rabbit, always fearing pursuit, has ear passages turned to the rear; and so the horse, which must listen for the driver behind. Those animals that make no sound have no ears. Eyes reveal the same correlation of part to use. Animals that move fast have eyes, and only the slow are blind.[17] In the preface to his earlier catalogue of the Royal Society's museum (*Musaeum Regalis Societatis*, 1681) Grew demonstrated the same point with some observations on the horse. Since it needs large thick hooves and strong teeth, the "excrementitious parts of the blood" are diverted to their production, leaving the horse without horns. The ox, on the other hand, since it needs horns, has fewer teeth and thinner hooves, providence making use of the blood to produce parts most suitable to each animal. Because of its peculiar teeth the horse needs its long upper lip, which serves it as the trunk serves the elephant and the hand serves man. Because it must have huge lungs, the horse must also have a broad chest, well-bowed ribs, and wide nostrils. Since it is pestered with flies, it must have a bushy tail to flick them off, while the thick-skinned donkey, little bothered by flies, neither needs nor has such a tail. In order to run swiftly the horse has pasterns to break the force of his weight, but the pasterns are not so long as to make each step laborious.[18] In all things the horse is perfectly fitted for the life it leads.

Grew discovered the harmony of parts in every creature reproduced on the broad canvas of the macrocosm, where all creatures conspire together for good. In his eyes, evil is far outweighed in the universal scales. If monsters are born, they are one in a thousand; and even these monsters are not without purpose, "serving to assist us in observing and valuing all those which are

17. Ibid., pp. 23–30.

18. Nehemiah Grew, *Musaeum Regalis Societatis. Or a Catalogue and Description of the Natural and Artificial Rarities Belonging to the Royal Society and Preserved at Gresham College* (London, 1681), preface.

regular."[19] Diseases upset our health, to be sure, but we are naturally fitted to throw them off. The most useful beasts and plants are the most plentiful. "A crane, which is scurvy meat, lays and hatches but two eggs in a year; and the alka and divers other seafowls lay but one. But the pheasant and partridge, both excellent meat, and more within our reach, lay and hatch fifteen or twenty together."[20] How did such harmony come to pass, Grew asked finally; could it be the result of chance? Did the eye make light by which to see, or the ear sound to hear? The answer was already present in the question. Only perfect intelligence, far above the operation of matter and chance, could have created such a universe of mutual cooperation.

Grew's procedure involved a gross misuse of the scientific method. While seemingly he was basing his argument on observed empirical facts, in actuality he was neither inducing a conclusion from a body of observations nor employing systematic observation to check a theory. He was selecting facts to support a conclusion that he already held. Like Charleton he did not consider the unpleasant aspects of nature, the monstrous births, the diseases, which he dismissed. Instead of including them in his interpretation of nature he reasoned them away. Those facts that he used were forced into an arbitrary pattern without the author's proving that they really fitted. Any conclusion was possible and facts were available to support any preconception when he could decide that the Creator turned the horse's ears toward the rear in order for it to hear the driver's commands.

The current of teleology, present in all of the virtuosi, reached its crest in the work of John Ray. *The Wisdom of God Manifested in the Works of the Creation* (1691), a book of great popularity, covered much the same ground that Grew went over, but covered it more thoroughly. Ray stated his purpose in the preface to the first edition. "The particulars of this discourse," he wrote, "serve not only to demonstrate the being of a Deity, but also to illustrate some of His principal attributes, as namely His infinite power and wisdom. The vast multitude of creatures . . . are effects and proofs of His almighty power. . . . The admirable contrivance

19. Grew, *Cosmologia Sacra,* p. 103.
20. Ibid., p. 99.

of all and each of them, the adapting all parts of animals to their several uses, the provision that is made for their sustenance . . . and lastly their mutual subserviency to each other and unanimous conspiring to promote and carry on the public good are evident demonstrations of His sovereign wisdom."[21] Fulfilling the outline in the preface, the book unfolded the intricately woven fabric of divine harmony.

First it was necessary to deal with Descartes' objection that because man is unable to discover the Creator's ends, philosophy must exclude the search for final causes. Ray asserted in reply that the purposes of some things are obvious and cannot be mistaken. The eye—he used the standard example—is so clearly meant for seeing that it is absurd to maintain that we do not know its purpose. Two further arguments that he advanced for final causes reveal his true motive. How, he inquired, can man praise God for the use of his limbs and senses and for those creatures that supply his sustenance if he does not know that they were intended for those uses? How can man prove the existence of God if the denial of final causes destroys the best demonstration for His existence?[22] Ray took up his study with the determination that nature must reveal God.

The Wisdom of God Manifested in the Works of the Creation organized its discussion in imitation of nature's organization. After some general statements favoring divine creation and denying chance and a word of approval for the atomic hypothesis, it first treated inanimate bodies, heavenly and mundane; and mounting the scale of being, it surveyed vegetable life and animal life. The second half of the book devoted itself mostly to the analysis of two particular creatures, the earth, considered as a physical body, and man, the highest form of animal life. In the solemn order of the heavens Ray found evidence of the Creator. Every body moves uniformly, both rotating on its own axis and revolving around another center. Exact uniformity of motion would be impossible if providence did not overrule chance, preserving the movement of bodies from internal change and external accident. As Ray proceeded, it became apparent that he was praising not

21. Ray, *Wisdom of God* (1st ed. London, 1691), preface.
22. Ibid. (4th ed.), pp. 42–4.

the heavenly order then known but the perfect heavenly order that he believed God had created and man would ultimately recognize. The more astronomers removed the irregularities of the old Ptolemaic system, the more they conformed to nature, and Ray had no doubt that the remaining irregularities would be banished when the true course of nature was discovered. To the extent that the hypotheses of astronomers are "more simple and conformable to reason," he affirmed, "by so much do they give a better account of the heavenly motions."[23] The same faith in the rational order of nature underlies the two basic axioms of natural philosophy that he delivered: "*Natura non facit circuitous*, Nature does not fetch a compass when it may proceed in a straight line; and *Natura nec abundat in superfluis, nec deficit in necessariis*, Nature abounds not in what is superfluous, neither is deficient in what is necessary."[24]

When Ray discussed the earth, he demonstrated how well such principles could be used to illustrate the wisdom of God. If nature has nothing superfluous and lacks nothing necessary, then all must be exactly right. The earth, for example, is a sphere, the most solid and compact body, and also the most convenient. If the earth were angular, parts of it would be huge mountains unfit for habitation. On the other hand if the earth were a perfect sphere, it would be covered entirely with water. As it is, the proportion of ocean to land is perfectly balanced to allow enough evaporation from the sea for the ground to be watered sufficiently with rain. We are fortunate in having what mountains there are, since they precipitate rain and provide us with mountain plants and animals. The rotation of the earth gives all of its surface the alternation of day and night, and the inclination of its axis produces the changes of seasons. Indeed just the right inclination ensures that the earth as a whole shall receive the greatest possible benefit from seasonal variations.[25]

In the smaller parts of creation Ray found the same felicitous adjustment of inanimate bodies to human needs. Employing the Aristotelian system of four elements to conform to popular usage,

23. Ibid., p. 72.
24. Ibid., p. 73.
25. Ibid., pp. 219–55.

he illustrated how necessary man finds fire, air, water, and earth. Fire not only warms us and cooks our food, it smelts the metals we use. Air sustains life. As for water, the seas provide fish; streams water the land and delight the eye. From the earth we receive the food that supports life. The elements work in harmony, moreover, reinforcing their utility by mutual cooperation. The winds are so ordered that they blow mostly from sea to land, bringing vapors that condense into rain. Excess rain may cause floods, but floods are useful in the end because they carry down dirt from the mountains and spread it on the plains. Even the size of raindrops is just right. If rain descended in a solid stream it would erode the ground, wash away plants, overturn houses, and greatly incommode if not suffocate animals. There was nothing apparently that did not reveal the goodness of the Father, and Ray could exclaim with justice, "O the depth of the riches both of the wisdom and knowledge of God." [26]

Whereas Nehemiah Grew employed teleology with some naiveté, unconscious evidently of the objections that could be raised against his arguments, Ray was aware of the possible weaknesses; and he paused several times to consider them. It might be maintained that the uses of things were not designed by nature, but that man accommodates things to his use.

> To which I answer . . . that the several useful dependencies of this kind (*viz.* of stones, timber, and metals for building of houses or ships, the magnet for navigation, etc., fire for melting of metals and forging of instruments for the purposes mentioned) we only find, not make them. For whether we think of it or no, it is, for example, manifest that fuel is good to continue fire, and fire to melt metals, and metals to make instruments to build ships and houses, and so on. Wherefore it being true that there is such a subordinate usefulness in the things themselves that are made to our hand, it is but reason in us to impute it to such a cause as was aware of the usefulness and serviceableness of its own works. To which I shall add that since we find materials so fit to serve all the necessities and conveniencies and to exercise and employ

26. Ibid., pp. 79–103.

> the wit and industry of an intelligent and active being, and since there is such an one created that is endued with skill and ability to use them, and which by their help is enabled to rule over and subdue all inferior creatures, but without them had been left necessitous, helpless, and obnoxious to injuries above any other, and since the omniscient Creator could not but know all the uses to which they might and would be employed by man, to them that acknowledge the being of a Deity it is little less than a demonstration that they were created intentionally, I do not say only, for those uses.[27]

The central issue at stake in Ray's argument seems to be the concept of instantaneous formation. When he maintained that the creation reveals the Creator, he was attacking the Epicurean position, which held that the universe is a chance concurrence of atoms. Any notion of evolution was unknown of course, and either position assumed that the world suddenly came to exist where it had not been before. Sun, sea, wind, and rain all began at once, and with them plants requiring about as much sun and rain as there are. Man, who is equipped to live and thrive under these conditions, appeared full blown. Given the assumptions, the rejection of the notion that chance could have produced the universe with its manifest mutual adjustment and cooperation between parts follows easily enough. The 17th century may be excused for leaving evolution to a later age; it discovered enough. Without a concept of evolution the virtuosi could think of life only in the terms in which they found it. The "creational" mentality, if it may be so called, was common to all of the virtuosi; there was no visible alternative. In starting from the premise that nature came into existence fully developed, they were only a step removed from the conclusion that God created it.

The creational mentality becomes more obvious when Ray's book moves on to the discussion of living creatures. In all of them he discovered a complex multitude of parts contributing to the existence and welfare of the whole. From the ground the roots of plants draw nourishment which the fibers carry and the leaves

27. Ibid., pp. 187–8.

elaborate. Every plant produces the seeds necessary for propagation, and nature protects the seeds with built-in defenses. Individual plants have special parts for special needs, such as thorns to prevent animals from eating them. Like inanimate bodies they are useful to man. Wheat, the most edible of plants, is also among the most hardy; it can endure extreme temperatures, and one seed will reproduce itself a hundred times.[28]

Among animals the harmony is maintained while the intricacy of parts is increased. To ensure the propagation of species nature has contrived a variety of special reproductive organs, and even timorous birds are equipped with a fierce instinct to defend their young. As wide in variety are the means of self-protection; some animals have strength, some hard shells, some merely speed. The sheep, which has neither strength nor speed, has been made useful to man so that he will protect it. Weak animals and birds breed often or have multiple young, while the birds and beasts of prey breed seldom and singly. As each species is equipped with the parts necessary for its mode of life, so birds in general have the special muscles needed for flight, carnivorous birds have hooked beaks, and water birds have downy breasts for warmth. Swallows, which live on flying bugs, are fitted for swift flight and have wide mouths, and woodpeckers have strong bills and long sticky tongues.[29]

The correspondence between Ray's material and Darwin's is as obvious as the difference of their conclusions. Both knew fully the specialization among species. What was proof of omniscient design to Ray became in the hands of Darwin evidence of evolution. Ray recognized several possible criticisms of his conclusions, criticisms that evolution was later to reinforce. There might be objections against the argument that proceeded from the ingenious contrivance of parts to the Creator. All parts are necessary to existence, it could be said; without them the animals would not exist at all. It might be claimed that nature has produced innumerable species in the course of time, and that those fitted to survive have survived. Couched in these terms, the objection, while it touched on arguments related to evolution, assumed the instantaneous

28. Ibid., pp. 116–32.
29. Ibid., pp. 132–87.

production of creatures by some means. Ray's reply based itself on the same premise. Many beasts have parts that are superfluous to mere existence even though they add to the convenience of life, he asserted. We do not really need two eyes, two nostrils, or two ears. We do not need nails, although they make life more pleasant by protecting tender areas. If the objection is valid, chance should have produced some one-eyed men or some men without nails, but there is no evidence of such either living or dead. Needs, Ray continued, existed before the parts that filled them. That is to say, God saw the needs and created the parts to satisfy them. As evidence he pointed to climbing plants. Only those plants with weak stalks have tendrils to pull themselves up. If chance brings forth everything, some plants with strong stems should have tendrils. Again, the armadillo and the hedgehog, which protect themselves by rolling up into balls, have special muscles to aid them, but no other animal has such muscles.[30] In an age that did not know the concept of evolution Ray's arguments must have carried considerable force. His knowledge of nature was wide and profound, and he marshalled his evidence with skill and insight. How completely the idea of evolution was to destroy his position!

More than the mere concept of evolution separated Darwin from Ray, however. A whole attitude toward nature, far more than a hypothesis, was involved in their difference. In Boyle a profound conviction of the benevolence of nature has already been remarked. It is necessary to recognize the same conviction in Ray in order to understand how the struggle for existence, implicit in his material, could not emerge into open day. Ray could point out the skillful defenses by which nature protects her weaklings, but he could not imagine peaceful creatures competing with each other to defend themselves from violence and to survive. Ray could describe the weapons of nature's destroyers, but he could not think of them as rivals for the limited supply of prey. He was blind to the implied contrast between the defenses of the weak and the weapons of the strong, both of which cannot be fully effective. Bred and trained as he was in the Christian religion in an age that had rejected the degeneration of nature, Ray

30. Ibid., pp. 412–20.

was incapable of picturing nature as the scene of continual violence. His God, he was sure, had created things for survival; and he could see only benevolence in the instruments that effected survival.

The virtuosi emerge as the mutual product of both empiricism and rationalism. Empirical in approach, they went straight to nature to observe the facts as they were. When it came to interpretation, however, they fitted the facts into a framework that stood on an a priori foundation; and major premises in their philosophy of nature were taken from the Christian tradition. The very concepts of natural law and natural order were inherited from centuries of Christian thought. Just as their reasoning on spiritual questions, such as the immortality of the soul, started from the assumption that creation was a rational order, so their reasoning in natural philosophy proceeded from the same basis and they were determined to find natural laws before they actually did. Their "creational" mentality, which unconsciously assumed that the cosmos had been formed or created fully developed, led them to attribute prior design to every element, where the 20th century sees development and evolution. Their conviction that God had created peaceful harmony focused their attention on the beauties of nature while they missed or ignored whatever there was of squalor and violence. Thus religion influenced their conception of nature. It was no wonder that the virtuosi discovered the glory of God in His works; they were looking into a mirror that reflected their own minds.

CHAPTER 4

Divine Providence and Natural Law

The heathen absurdly dreamed that all effects are inevitably produced by the conspiracy and coefficiency of natural causes . . . The more intelligent Christian proves . . . that the precise and opportune contingency of every individual event proceeds from the influence of this providence . . .

Charleton, *The Darkness of Atheism Dispelled by the Light of Nature*

And as it more recommends the skill of an engineer to contrive an elaborate engine so as that there should need nothing to reach his ends in it but the contrivance of parts devoid of understanding . . . so it more sets off the wisdom of God in the fabric of the universe that He can make so vast a machine perform all those many things, which He designed it should, by the mere contrivance of brute matter managed by certain laws of local motion . . .

Boyle, *A Free Inquiry into the Vulgarly Received Notion of Nature*

DESPITE the reverent attitude with which the virtuosi approached their studies and their concern to prove that the conclusions of natural philosophy did not challenge the claims of religion, they accepted and elaborated a conception of nature which posed problems for several traditional doctrines of Christianity. With few exceptions the virtuosi adhered to the atomic or mechanical hypothesis. In their opinion it was the true Christian philosophy, the conception of nature whereby the existence of God was most clearly revealed. According to Robert Boyle, the Aristotelian philosophy, which had dominated thought in the Middle Ages and still reigned in the universities, attributed all phenomena to some supposed being, which it called nature. It was likely to lead to the conclusion that nature is independent of God, whereas the mechanical conception clearly pointed to a Creator Who had made the machine. Nevertheless the mechanical theory of nature forced the virtuosi who accepted it to modify their interpretation of some Christian doctrines. While they were protesting vigorously that natural philosophy was in harmony with Christianity, they were quietly and perhaps unconsciously altering Christianity to meet their own definition. Their sincerity is not in question; they certainly considered that their version of Christianity was the true one. But in their version doctrines grouped around the idea of providence were changed from their traditional meaning. Thus notwithstanding their piety the virtuosi introduced significant innovations into Christianity.

To the virtuosi the Aristotelian philosophy, which they rejected in favor of the mechanical conception of nature, seemed empty and meaningless. It attempted to explain natural phenomena by the action of substantial forms; the virtuosi complained that explanations in such terms only glossed problems over with words that meant nothing, obscuring instead of revealing the true causes of things. Through their works a constant chant of protest swells against the counterfeit philosophy which was obstructing the progress of science. To satisfy them an explanation had to be couched in terms which described the supposed operation of nature's mechanical parts. Robert Hooke echoed the protest against Aristotle in the preface to his *Micrographia* (1665) when he spoke of the aid that the microscope would bring to human senses.

> It seems not improbable but that by these helps the subtlety of the composition of bodies, the structure of their parts, the various texture of their matter, the instruments and manner of their inward motions, and all the other possible appearances of things, may come to be more fully discovered; all which the ancient Peripatetics were content to comprehend in two general and (unless further explained) useless words of matter and form. From whence there may arise many admirable advantages towards the increase of the operative and the mechanical knowledge to which this age seems so much inclined, because we may perhaps be enabled to discern all the secret workings of nature, almost in the same manner as we do those that are the productions of

art, and are managed by wheels and engines and springs that were devised by human wit.[1]

The difference in outlook between the two schools of thought can be seen in the words of two other virtuosi. Thomas Sydenham, one of the few virtuosi who remained an Aristotelian, declared in his essay on rational theology that all creatures "are put under laws, by which they are determined to such or such operations suitable to the ends of their several beings . . ."[2] In Sydenham's opinion nature is a hierarchy of creatures, each striving to realize the end for which it was created, governed by a law but not a mechanical law. Opposed to his conception are the words of John Locke. Sensible objects in his view are made up of "inconceivably small bodies, or atoms, out of whose various combinations bigger moleculae are made; and so, by a greater and greater composition, bigger bodies; and out of these the whole material world is constituted. By the figure, bulk, texture, and motion of these small and insensible corpuscles all the phenomena of bodies may be explained."[3] Locke's conception was the one accepted by nearly all of the virtuosi. It too held that the actions of bodies obey laws, not laws suitable to the ends of their several beings, but mechanical laws which do not include the idea of striving for ends. In short, nature was looked upon as a machine made up of unconscious material parts controlled by external mechanical forces.

There was no standard mechanical conception of nature accepted by all of the virtuosi. They were building the conception, not merely accepting it, and no two men agreed on all of the details. Some men interested themselves in scientific discoveries without worrying about philosophical foundations at all. Some created unstable amalgams of Aristotelianism and atomism. Among those who frankly embraced, or tried to embrace, the mechanical philosophy, non-mechanical elements kept creeping back under sundry disguises when difficult phenomena had to be explained. There were nearly as many mechanical conceptions of

1. Hooke, *Micrographia*, preface.

2. Thomas Sydenham, *The Works of Thomas Sydenham, M.D.*, trans. R. G. Latham (2 vols. London, 1848–50), 2, 309.

3. *The Works of John Locke* (9th ed. 9 vols. London, 1794), 2, 440.

nature as there were virtuosi, and no two presented exactly the same problems. Nevertheless there was general agreement on the goal in mind. With varying success according to their different intellectual powers the virtuosi tried to construct a theory picturing nature as a machine running by itself without external aids, a machine which human science could study and comprehend.

Perhaps the best example of the virtuosi's conception of nature is found in the writings of Robert Boyle (1627–91), the most influential publicist of the mechanical philosophy in England. Boyle devoted all of his philosophical works to substantiating the mechanical hypothesis, and the popularity of the clock analogy in explaining the operation of nature was due largely to his writings. Borrowing the figure from Cicero, he polished it and developed it and passed it on to posterity as the representation of the 17th century's conception of nature. Nature was compared to a huge machine composed in turn of lesser machines. The complex clock on the Cathedral in Strasbourg was the example that best illustrated the point, and Boyle cited it continually. Its movements were so regular that it seemed to be an intelligent creature; indeed Boyle reported that when Europeans took a clock to the East, the Orientals thought that it was alive. Each part of the Strasbourg clock performed its function as though it understood the purpose of its movement, yet the clock was dull lifeless matter, unconscious of what it did. Boyle maintained that nature is like the clock. Inert matter moves and mechanically causes phenomena which seem to be the work of intelligence. To explain why liquids rise in a closed tube the old philosophy had said that nature abhors a vacuum, which apparently meant that liquids can perceive a vacuum and move to fill it. The explanation implied that water is animate. The advocates of the mechanical hypothesis replied, with experimental evidence as proof, that unconscious mechanical pressure is the cause of the phenomenon. As there is no intelligence in liquids to abhor a vacuum, so there is no intelligence in any material body to comprehend its own motion. Like the pieces in the clock, bodies move unknowingly in their orbits, pushing their neighbors and being pushed by them, unconsciously doing their allotted work.

Intelligence was at work in the original creation, shaping and placing the pieces with great skill; since the creation nature has run its blind mechanical course.

Obviously the mechanical idea of nature was closely bound up with atomism. The mechanical hypothesis required some fundamental particles to act as the primary parts of the clock. Boyle was a convinced atomist who brought extensive evidence from his chemical laboratory to support the contention that matter is composed of fundamental, irreducible particles. In his literal imagination atoms were transformed into gears and levers which were mentally pictured as running the basic parts of the cosmic clock. When he spoke of atmospheric pressure as the spring in the air, he meant real springs, the particles of air being thought of as little coils which contract and expand. Thus the Strasbourg clock was a precise image of Boyle's conception of nature. His world was eminently picturable, with atoms, tiny levers and wheels, pushing and pulling and turning each other to make the universe go.

The mechanical conception of nature as Boyle developed it was the one accepted, in its major outlines at least, by nearly all of the contemporary and younger virtuosi. Not that they were slavish imitators; everyone had his own particular ideas, but nearly all of them agreed that physical reality consists of small inert material particles moved and controlled by mechanical impulses. General laws of motion that describe the movement of bodies in exact mathematical terms occupied a greater place in later minds, and Newton's concept of force added a refinement, but the general outlines of the picture remained unchanged. While Boyle did not invent the conception, his writings and his experiments did much to make it credible, as it had not been in earlier writers such as Sir Kenelm Digby and Walter Charleton. His place in the history of science lies more perhaps in the field of publicity than of discovery. His books were widely read, as their frequent republication indicates; he was held in high esteem by the virtuosi, and his influence was great. That the virtuosi belonging to the generations that followed him accepted the atomic and mechanical hypothesis almost unanimously reflects in part the success of his effort.

Mechanical nature was not atheistic; Boyle easily brought the clock analogy to the service of theism. He liked to picture a man examining an intricate and complex clock. Could he conceivably think that the clock is a work of chance, as if someone had thrown down some scraps of metal and they had fallen together into a fine mechanism? Must he not conclude that the clock is the work of a skillful artificer? The clock analogy served very well in Boyle's contention that nature reveals her Creator.

The clock analogy also emphasized the challenge to the Christian doctrine of particular providence inherent in the mechanical conception of nature, for the operation of a clock is pre-ordained by its maker. In the Christian tradition the idea of providence had not been limited to the Almighty's general governance and maintenance of the universal order. In addition to upholding the universal order, God was supposed to watch over each of His creatures with fatherly care, cherishing and protecting, rewarding and punishing. Since the concept of providence had not been restricted to the welfare and salvation of the soul, it had implied constant activity by the Almighty or His agents in the operation of the physical world. For instance, recovery from a disease was often attributed to divine providence, meaning that God had altered the natural course that the disease would have taken had He not acted. Divine providence conceived in these terms is incompatible with the mechanical idea of nature, for divine intervention in nature would involve the violation of the same laws of motion that God Himself created. In the early years of the 18th century Leibniz stated the dilemma in its baldest terms. As part of his controversy with Newton, he exchanged a number of philosophical letters with Dr. Samuel Clarke, who defended Newton's position. Denying the Newtonian conclusion that God must correct aberrations in the cosmic machine from time to time, Leibniz attacked the idea that God intervenes in His creation at all. Newton, he charged, had conceived of God as a watchmaker who must wind his piece now and then, even clean and mend it. Leibniz pointed out that the more often God must mend His work, the more unskillful He must be. Divine interventions in nature would imply self-contradiction on the part of the Almighty Himself. Leibniz' argument applies equally well to the traditional doctrine of particular providence. A concept of general providence, whereby the Creator preserves the whole system and maintains its laws, and a doctrine of particular providence that confines itself to spiritual welfare and salvation are compatible with the mechanical hypothesis, but the reign of mechanical laws of motion outlaws the idea of particular divine protection for the material bodies of individual creatures. If the notion of providence is limited in its meaning, the reality of miracles, which were both recorded in the Bible and bound up closely with Christian tradition, will be renounced.

In actual fact it is wrong to lay the troubles besetting the doctrine of providence to the mechanical theory alone. The challenge to providence was no more inherent in the mechanical philosophy of nature than in the peripatetic. Aristotle had rejected the idea of providence; and Thomas Sydenham, an Aristotelian virtuoso, wrestled with the same difficulties in his treatment of providence that troubled the virtuosi supporting the mechanical hypothesis. However, the Christian philosophers of the centuries preceding the 17th had not been concerned primarily with the problems of physical and efficient causation, nor had they been driven by the urge to predict phenomena and control nature. With their concept of the natural order as a subordinate aspect of God's providential plan they had adequately reconciled providence with natural law. With their concept of divine prevision in the eternal present they had explained how God could adjust agents to His providential ends without violating the natural order. With their concept that God's concurrence is necessary for the operation of contingent agents they had demonstrated how providential activity could be an ever-present reality. During the 17th century physical causation, prediction, and control became pressing intellectual problems. No longer was nature considered as an element in the providential plan. It might be identified with general providence because God was thought to have created it with the maximum inherent beneficence and goodness, but nature itself was taken to be an order of static, immutable, and impersonal laws. Material causes, once created, were supposed to act by their own necessity in accordance with natural laws. Within this intellectual framework the reconciliation of the traditional doctrine of particular providence with the natural order assumed new difficulties; and it was the mechanical conception of nature, one element of the scientific movement of the 17th century, which clashed with the Christian idea of providence.

Beyond the challenge to providence which is present in any well developed conception of nature, the metaphysics adopted by the virtuosi contains a problem peculiar to itself. The conception of mechanical nature involves the exclusion of spirit from the physical realm. God is thought of as the Creator Who built the machine originally and set it in motion; but once created, nature is an autonomous material order made up of senseless atoms blindly moved by other unconscious particles. Mechanical nature does not leave any function for God in His creation, and it excludes the spiritual soul of man as well. Although the human soul may observe the order and even at times act as a sort of mechanical force in moving bits of matter, it is not an integral part of the physical world. Mechanical nature can function equally well if the human soul is present or not. In a word, the virtuosi stripped away the defenses against materialism, although they were decidedly not materialists themselves.

While the challenge to certain Christian doctrines inherent in the mechanical conception of nature can be pointed out now, the problem was seen less clearly in the light of the 17th century. The virtuosi themselves refused to believe that the mechanical hypothesis could call Christianity into doubt. Although they recognized that their discoveries destroyed a host of superstitions that had become attached to Christianity in one way or another, they could not see any conflict with what they considered to be essential doctrines. Their approach to the problem of providence reveals a good deal about the religious attitude of the virtuosi. Instead of concluding that the idea of providence is untenable, they tried to preserve it by some formula that would reconcile it adequately with their conception of nature. By reinterpreting the concept of providence, they thought that they had saved the doctrine and demonstrated the harmony of natural science and religion. The apparent conflict between science and religion over providence was further softened by another characteristic of the virtuosi. They did not always draw the full implications of their theories, and they were more ready to compromise a scientific principle than to surrender an essential religious belief. The reality of miracles is a case in point. Their Protestant convictions as well as their scientific opinions assured them that the age of miracles was past; the ordered course of nature precluded miraculous eruptions in their day. Most of them would not, however, accept the obvious corollary of their scientific principles that miracles seventeen centuries earlier were just as impossible. The Bible recorded miracles, and they regarded the Bible as the inspired foundation of the Christian religion. To question the biblical miracles was to question all of Christianity, and this the virtuosi would not do. In the problem of miracles the depth of their religious convictions can be tested, for here the virtuosi turned aside from the conclusion of science to accept a conflicting doctrine of religion. Only at the end of the century did a few men begin to question the biblical miracles. The virtuosi, then, did not look upon the problem of providence as a conflict of scientific conclusions with Christian beliefs. Since they did not regard science as a yardstick by which to measure Christianity, their procedure was to find a ground of reconciliation which satisfied both aspects of what they considered to be one truth.

Some of the early virtuosi did not recognize the problem at all. Such men as the two Baconian reformers, Samuel Hartlib (ca. 1600–62) and John Beale (1603–83), expressed great interest in individual scientific discoveries but did not integrate them into a theory of nature. Evidently they never considered that the conclusions of science might conflict either with Christianity in general or with the doctrine of providence in particular. The same may be said of another virtuoso of the same generation, Dr. John Wilkins (1614–72), although he did discuss providence enough to bring out the two antitheses that had to be reconciled. In *A Discourse Concerning the Beauty of Providence,* which he published in 1649, Wilkins did not make any compromise with the realm of physical necessity. Since his attention in the *Discourse* focused itself on the meaning of providence for the individual, he was not concerned with the secondary question of material causality. He was able to assert without hesitation that there is nothing in the world so great that it is not under God's power, nothing so small that it is not within His care. Returning to the consideration of providence again in a context dealing with the natural order, in the posthumous *Principles and Duties of Natural Religion* (1675), Wilkins affirmed the reality of providential care once more—particular providence, not general providence, as he specifically stated.[4] In developing the fundamentals of natural religion, he contended not solely for the existence of God but also for the existence of a God Who can alter the course of nature. This was half of the problem—the reality of God's power to rule His creation. Had Wilkins not been concerned with natural philosophy, his position would have been simple. Since the Almighty Father places the care of His children before the consistency of natural laws, He exercises His immediate rule of all things to guide and protect them. Wilkins was concerned with natural philosophy, however, and in the same treatise on natural religion he declared that all material things move by necessary laws.[5] This was the other half of the problem—the reign of natural laws in the creation, a contradiction, if rigorously pursued, to the Almighty's power of immediate governance. When Wilkins discussed miracles later in the book, the contradiction stood out more clearly. Miracles, he said, were limited to the days when Christianity was young and in need of supernatural testimony. Now the age of miracles was past and properly so, "it being not reasonable to think that the universal laws of nature, by which things are to be regularly guided in their natural course, should frequently or upon every little occasion be violated or disordered."[6] Wilkins left the problem of providence apparently without recognizing the inherent conflict between the two positions he took. Undoubtedly the disparity was obscured in his thought by the overriding faith in divine goodness shared by all of the virtuosi. Believing implicitly in the benevolence of the original creation, he never stopped to question whether the impersonal rule of natural law might blight the welfare of indi-

4. John Wilkins, *Of the Principles and Duties of Natural Religion* (London, 1675), p. 130.
5. Ibid., p. 108.
6. Ibid., p. 402.

vidual beings. Although he offered nothing by way of solution, he did state the two opposites, to the reconciliation of which other virtuosi bent their efforts.

When the system builders took up the question, the latent conflict between the two positions stated by Wilkins came to the surface. Two other virtuosi of the early generation tried to reach some statement of providence that would resolve the antithesis. Sir Kenelm Digby (1603–65)—the magazine of all arts, as he was styled—and Dr. Walter Charleton (1619–1707), President of the Royal College of Physicians, were men of a different intellectual type from Hartlib, Beale, and Wilkins. Would-be philosophers who had fallen under the influence of Descartes and Gassendi, they tried to construct rational systems of nature which explained phenomena by mechanical causes. Both accepted the Cartesian dualism of matter and spirit; they severed the intimate connection between the two and left nature a machine operating independently of spiritual control. Since neither Digby nor Charleton reached great philosophic stature, their systems are of little interest in themselves, beyond their illustration of the influence that the mechanical hypothesis exercised over the minds of the virtuosi. Their ideas of providence, however, are of some significance. They both expressed themselves on the meaning of providence, and their declarations were framed with the mechanical hypothesis in mind. Their antithetical opinions indicated the two possible directions in which attempts to resolve the dilemma could move.

Digby's stand foreshadowed the position that the virtuosi tended more and more to accept—that providence is real, but that providence is the Creator's original plan alone. In the beginning God built the machine of nature, foresaw all of its effects, and fashioned its parts to perform what He wanted to be accomplished. "I believe that all causes are so immediately chained to their effects," Digby wrote in his observations on Sir Thomas Browne, "as if a perfect knowing nature get hold but of one link, it will drive the entire series or pedigree of the whole to each utmost end . . . so that in truth there is no fortuitousness or contingency of things, in respect of themselves, but only in respect of us, that are ignorant of their certain and necessary causes." Divine providence is such a chain of causes and not a secret, invisible, and mystical process which does not fall under the search or cognizance of a prudent investigation.[7] To some extent Digby's theory approximated the Thomist position, which held that nothing happens by chance in the eyes of omniscient God. God sees everything at once in the eternal present according to St. Thomas; He knows for all time all of the chance encounters that will cross the map of history. To the Thomist, however, the great spectacle of providence is hidden forever from human eyes, which cannot see at a glance the whole of nature compressed into the eternal present. Digby on the other hand thought that providence operates by a mechanical necessity inherent in the original construction of the machine. Without mentioning the eternal present, he expounded providence as a succession of mechanical operations much like the movement of a clock ticking off the seconds one after another. The determinate succession would be quite comprehensible to man once he understood the machine. Whoever grasps one link in the chain, he wrote to an unnamed lady in France, "and knows how to use it, to draw it skillfully and pass it adroitly through his hands, will easily reach both ends." [8] In Digby's theory God is reduced to the role of original Creator, Who made the machine and foresaw its actions but has not bothered further since it began to run.

Charleton took up an alternative solution. In the *Darkness of Atheism* (1652) he argued that divine omnipotence still governs the world which it created. He classified as implicit atheists those who admit a God and then deny Him the governance and conservation of the world.[9] In Charleton's opinion there is no room in the universe for chance. The voracity of beasts, the ravages of weather, the onslaughts of disease, "all these are the regular effects of God's general providence, and have their causes, times, and finalities preordained and inscribed in the diary of Fate, to whose prescience nothing is contingent." [10] At first blush

7. Kenelm Digby, *Observations upon Religio Medici,* published with *Religio Medici* (Oxford, the Clarendon Press, 1909), p. 10.

8. Digby to Madame ———, 15/25 September (probably late 1630's); British Museum, Harleian MS 4,153, fol. 109.

9. Page 4.

10. Ibid., pp. 92–3.

his analysis appears identical with Digby's, but Charleton went on to declare that all things are not the inexorable result of natural laws. Against the stoic conception of determinism, which held that all effects are inevitably produced by the operation of natural causes, he asserted the Christian idea of determinism. Christians believe that although God foresees all events, the influence of His providence in disposing contingent natural causes is necessary to produce them. As God made nature, "so He can alter her and tune all her strings to a concord with His will . . ."[11] Charleton declared that the stoic position strikes at the cardinal attribute of the divine nature, omnipotence, "by coercing His infinite and arbitrary activity with the definite laws of second causes, and denying Him the prerogative of absolute superiority to His mechanical vicegerent, or (rather) instrument, nature."[12] The notion of an absolutely determined natural order impugns God's power to alter His own laws, "chaining up His arms in the adamantine fetters of Destiny."[13]

To prove that God can transcend and reverse the order of nature Charleton pointed out that there had been miracles in the past. Citing three from the Bible—the deluge, the cessation of oracles and the circumscription of Satan's power after the coming of Christ, and the eclipse of the sun upon His death—the doctor asserted that they could not possibly have been natural events; they must therefore have been miracles wrought by God. In contrast to most of the virtuosi he contended that God might do the same in the future whenever He so decided; that is, he believed that the age of miracles was not past.[14]

Thus to state the case was to save the doctrine of providence with disarming ease. Unfortunately Charleton's argument also destroyed any meaningful concept of natural law and natural order. Carried to its logical conclusion—as it seemed to be in certain passages—it would remove every action from the realm of natural causation to the sphere of the supernatural. It would have derogatory overtones in religion as well, making the creation

11. Ibid., p. 125.
12. Ibid., p. 217.
13. Ibid., p. 329.
14. Ibid., pp. 137–52.

of natural laws an act surprisingly imperfect for an omniscient God. In another discussion Charleton saw the difficulty. When he was trying to prove the immortality of the soul, he held that no material cause can destroy an immaterial substance. Hence only God can dissolve an existent soul. But God has pronounced His creation good. Although He has the power to annihilate, it would contradict His wisdom to do so; "for supposing (as we ought) that God does nothing contrary to the established laws and decreed order of nature, and that this general state of things does continue still the same, which His wisdom at first instituted, it evidently follows that what He has once made incorporeal shall persevere to be the same to all eternity." [15] Charleton did not apply this argument to his own statement of divine providence. He continued to stand on his original assertion that God reserves the power to govern His creation and to intervene in its operation.

Later virtuosi tried to strike a balance between the conclusions of Digby and Charleton. The generation following the two men generally accepted Charleton's solution, but restricted its application by asserting that God will intervene only on rare and important occasions. They also tended more and more to interpret providence as the benevolence of the original creation instead of immediate fatherly care; that is, general providence was displacing particular providence. As scientific investigation laid bare the operation and scope of material causes, the virtuosi found it increasingly difficult blandly to assert that God might step in at any moment to reshape the course of nature. The conception of the Almighty was gradually changed as He came to be pictured—as He was in the writings of Digby—in the role of Creator alone.

Robert Boyle, who developed the mechanical conception of nature most fully, was responsible also for the most penetrating examination of its meaning for the doctrine of providence. When he surveyed the question in his early philosophical works, Boyle thought that Charleton's solution was adequate. He asserted that God cannot be bound by His own creation; the strong right arm of the Almighty is still free to rule His creatures. As he pondered

15. Walter Charleton, *The Immortality of the Human Soul, Demonstrated by the Light of Nature* (London, 1657), p. 81.

the question more deeply, however, he recognized that such a solution only scratched the surface: as he said, "it much more tends to the illustration of God's wisdom to have so framed things at first that there can seldom or never need any extraordinary interposition of His power." Reverting to his favorite simile, Boyle expounded his opinion by comparing God to an engineer. An engineer exhibits more skill in designing a machine that will do whatever job he has in mind by itself than in making one that requires an operator. In the same way God displays His wisdom more in creating a universe that fulfills all of the purposes that He intends by the operations of brute matter alone than in fashioning one that needs an intelligent overseer to regulate and control its parts.[16] The passage occurs in a work entitled *A Free Inquiry into the Vulgarly Received Notion of Nature,* in which Boyle presented a fully elaborated discussion of providence.

The "vulgarly received notion of nature" against which Boyle was contending was the concept of plastic nature, which some of the Cambridge Platonists considered essential to the doctrine of providence. On the one hand nature, as vicegerent of God, was said to act as the Almighty's agent in overruling the motions of matter from time to time; on the other hand nature, being less than God, was made responsible for the anomalous productions of nature which seem to impugn divine wisdom and justice. Thus the concept of plastic nature attempted to vindicate divine providence in two ways, providing God with an instrument through which He governs the universe and intervenes in its operations when necessary, and absolving God of responsibility for phenomena that seem to deny His goodness. Boyle's objection stemmed from a radically different idea of providence: he perceived the Almighty's goodness and power in the order and harmony of creation, his awareness of which exceptional events did not outweigh and anomalies did not upset. God demonstrated His care for the welfare of His creatures in the perfection of the original creative act. In His wisdom and goodness the Father so contrived the world "that even those creatures of His, who by their inanimate condition are not capable of intending to gratify me, should be as serviceable and useful to me as they would be if

16. *Works,* 5, 162.

they could and did design the being so."[17] Providence cannot therefore justly be denied because of a few anomalies, for such events and beings may serve divine ends beyond our comprehension. Since there are intelligent creatures in the world, one of God's intentions may be to show them His goodness in the variety of His works rather than in the perfection of each individual one. His wisdom is so far above ours that defects in our eyes may be perfections in His. The tendency of Boyle's remarks is clear. He did not really reply to the argument that plastic nature is the instrument by which God can overrule natural motions when necessary. In effect he ruled out the necessity of divine intervention and defined providence as the maintenance of the universal and benevolent order—that is to say, general providence.

Boyle's intuition of divine providence in the ordered harmony of nature can be seen in another of his works which was published in the same period, *Of the High Veneration Man's Intellect Owes to God* (1685). The vast complexity of the world, he argued, and the great variety of creatures in it are overwhelming evidence of the power and wisdom of God.

> For there being among these a stupendous number that may justly be looked upon as so many distinct engines and many of them very complicated ones too, as containing sundry subordinate ones, to know that all these as well as the rest of the mundane matter are every moment sustained, guided, and governed, according to their respective natures and with an exact regard to the catholic laws of the universe; to know, I say, that there is a Being that does this every where and every moment, and That manages all things without aberration or intermission, is a thing that, if we attentively reflect on, ought to produce in us for that supreme Being That can do this the highest wonder and the lowliest adoration.[18]

The order of nature, the unfailing rule of natural law over brute matter, dominated Boyle's imagination as no miracle could. The first and greatest miracle is the creation itself, an act of omnipo-

17. Ibid., p. 162.
18. Ibid., pp. 140–1.

tence which puts a mere extraordinary event to shame. The perception of a universe ruled by eternal immutable laws, however much it may have been an unconscious presupposition, was one of the aspects of nature that so excited the religious awe of the virtuosi. The vindication of natural law assumed greater importance in their eyes than the care and protection of any particular creature.

Boyle's discussion of providence was veering in a direction that had to exclude particular providence. He was not yet finished, however, nor was he ready to dispense with the notion of particular providence. Continuing with the statement that God had several ends in mind when He created the world, he suggested that although some may be knowable to us, others are not. Among those that we can discern are the manifestation of the glory of God, the utility of man, and the maintenance of the system of the world, including the preservation of particular creatures. The notion of particular providence reappeared, but Boyle immediately introduced two conditions which limit God's provision for the maintenance of particular creatures. The faculties that God gave them to effect their preservation are meant for the ordinary and usual course of events; they may prove a hindrance in unusual circumstances. A mother at childbirth, for instance, has plentiful milk; if the baby dies, the milk may become dangerous to her health. Furthermore, God places His care of individual and particular creatures below the welfare of the whole system, and particular creatures may have to suffer to preserve the whole. God subordinates "His care of their preservation and welfare to His care of maintaining the universal system and primitive scheme or contrivance of His works, and especially those catholic rules of motion and other grand laws, which He at first established among the portions of the mundane matter." [19] The limitations that Boyle made rendered his suggestion of particular providence a chimera and left him where he was before. In setting the preservation of the cosmic order above the fortune of individuals, he reduced divine providence to the original scheme of creation and destroyed any meaningful concept of particular providence. When God called the cosmos forth from

19. Ibid., p. 199.

the void, He exercised His benevolence in creating an order which provides for the greatest good of the greatest number. Having established a universal order charged with beneficence, the Almighty restricts His governance to maintaining it, preferring the occasional suffering of individual creatures to the violation of the created order that must accompany any action to mitigate their hardships. If Boyle had rested his case at this point, he would have allowed the doctrine of particular providence to disappear into the concept of general providence.

As a Christian, however, Boyle recognized that God had intervened in the creation to upset the normal course of nature. The Bible reported miracles, and Boyle neither could nor would depart from his belief in them. In the *Free Inquiry into the Vulgarly Received Notion of Nature* Boyle asserted that divine providence is often concerned with the actions of men. In the conduct of the greatest part of the universe, which is merely corporeal, God seldom causes a recession from the settled course of nature; yet where men are concerned,

> I think He has, not only sometimes by those signal and manifest interpositions we call miracles acted by a supernatural way, but as the sovereign Lord and Governor of the world does divers times (and perhaps oftener than mere philosophers imagine) give, by the intervention of rational minds, as well united as not united to human bodies, divers such determinations to the motion of parts in those bodies and of others which may be affected by them as by laws merely mechanical those parts of matter would not have had, by which motions, so determined, either salutary or fatal crises and many other things conducive to the welfare or detriment of men are produced.[20]

To the idea that providence can operate through the agency of free and rational man Boyle never returned. In addition to that suggestion he also affirmed, in the passage, that God intervenes, or has intervened, at times in the physical order to work actual miracles in His care for His creature man. For the present Boyle did not examine the question of miracles thoroughly; he merely

20. Ibid., pp. 215–16.

asserted that miracles were not fictitious, thus keeping the door ajar for a broader idea of providence.

At the end of the *Free Inquiry into the Vulgarly Received Notion of Nature* Boyle summed up his argument on providence. God freely created the world *ex nihilo* with several ends and purposes in mind, some of which we may know, some not. In the creation the preservation of the whole must be more important than the welfare of particular individuals.

> Upon these grounds, [he continued] if we set aside the consideration of miracles, as things supernatural, and of those instances wherein the providence of the great Rector of the universe and human affairs is pleased peculiarly to interpose, it may be rationally said that God . . . clearly discern[ed] what would happen in consequence of the laws by Him established in all the possible combinations of them . . . And that having, when all these things were in His prospect, settled among His corporeal works general and standing laws of motion suited to His most wise ends, it seems very congruous to His wisdom to prefer (unless in the newly accepted [*sic*] cases) catholic laws and higher ends before subordinate ones, and uniformity in His conduct before making changes in it according to every sort of particular emergencies; and consequently, not to recede from the general laws He at first most wisely established to comply with the appetites or the needs of particular creatures or to prevent some seeming irregularities (such as earthquakes, floods, famines, etc.) incommodious to them, which are no other than such as He foresaw would happen . . . and thought fit to ordain, or to permit, as not unsuitable to some or other of those wise ends which He may have in His all pervading view, Who . . . may have ends . . . divers of which, for aught we can tell or should presume, are known only to Himself; whence we may argue that several phenomena, which seem to us anomalous, may be very congruous or conducive to those secret ends, and therefore are unfit to be censured by us dim-sighted mortals.[21]

21. Ibid., pp. 251–2.

In Boyle's summary the notion of general providence triumphed completely, yet as a Christian he felt compelled to struggle against the conclusion and to introduce the passage with an exception which contradicted the principle. Unwilling to give up the idea of particular providence completely and unable to deny the biblical miracles, he forced himself to make miracles an exception to the general rule.

Among Boyle's papers in the Royal Society are a number of fragments on miracles which indicate that the problem gave him considerable trouble. The inspired word of God contained indubitable evidence that miracles had occurred; it was necessary to make allowance for them. As to the physical problem involved, it seemed clear that omnipotent God must have the power to overrule His own creation at any time. Miracles cannot be impossible for an omnipotent agent. In spite of what Boyle himself said about God's wisdom in establishing a machine that requires no intervention, he agreed that the laws can be suspended on some occasions. "I say . . . that God is a most free agent, and His divine wisdom does accompany all that He does in such a manner as not to impair His freedom but concur to accomplish the exertions or issues of it in the best manner that is possible." We cannot know all of God's ends, and it is a bit magisterial on our part to assume that He has none that may require that some of the laws established by Him should now and then "(though very rarely)" be controlled and receded from.[22] The argument, which reflects Boyle's Protestant training, that God has ends beyond human comprehension was a versatile tool. Whereas he used it earlier to contend that anomalies in nature are not necessarily violations of her order, here he employed it to prove that God is free to violate the natural order whenever He feels it to be necessary. That he added the parenthetical phrase "though very rarely" shows that he vaguely recognized the inconsistency.

Boyle's opinion about miracles stood in absolute contradiction to the rest of his thought. It was an artificial and arbitrary reconciliation of two positions that could not be reconciled satisfactorily. Obviously it worried him, and he returned to ponder it often, both in numerous papers that never saw publication and in

22. Royal Society, Boyle Papers, 7, fol. 113b.

the *Christian Virtuoso* (1690). Not wishing to open the door to popular credulity, he made it clear that miracles are rare, very rare. Only matters of overwhelming importance can lead God to violate the laws of His own creation, and in the history of the world only the promulgation of Christianity warranted such action. In one of his unpublished papers, "Some Considerations about Miracles as They Are Pleadable for the Christian Religion," Boyle stated that if he were restricted to one argument for the Christian religion, he would choose the miracles that have attested to it. The validity of the biblical revelation depends upon the miracles that prove it to be from God. Since Christianity is God's greatest legislative act relative to men, it is inconceivable that He would have promulgated it without sufficient signs to prove that it is His law. To inform by miracles, Boyle added, is a manner of teaching worthy of God, for men can teach by reason, but only God can work through omnipotence. Miracles are a proper way to appeal to men of all capacities; subtle arguments may convince philosophers, but Christianity is meant for all men.[23] With the foundation of Christianity, Boyle agreed, miracles ceased. In another paper, in which he tried to distinguish the truly miraculous from the spurious and superstitious, he advanced the idea that a miracle is known by the doctrine it teaches. The doctrine must be correct to prove that the miracle is real.[24] When he introduced such a criterion, Boyle gave his whole position away: miracles were to be judged by doctrines and doctrines were to be judged by miracles. He was caught within a closed circle from which there was no escape. Actually he did not want to escape from the circle. To take a straight line on miracles would have been to reject one of the two alternatives, while he preferred the inconsistency which allowed him to assert them both. Boyle's protracted yet unsuccessful wrestling with the problem indicates the ultimate incompatibility of the two extremes that he was trying to reconcile; but Christian virtuoso that he was, he could not bring himself to use science as the test of religion's validity. He was content to stop short of his logical conclusion and to maintain his belief in the Christian miracles and the doctrine

23. Ibid., 7, fols. 105–10.
24. Ibid., 7, fol. 120.

of particular providence. Even so, how little of the traditional notion of providence survived his reasoning. Only a few miracles remote in time confirmed God's power actively to direct and govern His creation.

Beyond the question of particular providence Boyle also feared the conclusion that the material realm is an autonomous order. Wishing to demonstrate that spiritual power is the dominant factor in nature, he worked over his natural philosophy, apparently looking for any province where God's continued participation in the creation is either possible or required. In the *Christian Virtuoso* he put forward two arguments. Local motion is adventitious to matter. It was first produced and is still maintained and preserved immediately by God. It may therefore be inferred that God concurs in the acts of each particular agent. Moreover, the rational soul, as an immaterial substance distinct and separate from the body, must be created specially and joined to each embryo at some point in its growth.[25] The second point, which Boyle mentioned in passing, did not occupy an important place in his philosophy; but the first, as he expanded it, clarified an ambiguous phrase which he often employed. Throughout his works Boyle spoke of the "ordinary and general concourse" of God in the operation of nature. The Almighty must continue to support the universe which He created, or it will dissolve into the void. Boyle never stopped to define the phrase, and its meaning evidently varied from passage to passage. At times he seemed to have no more than a passive acquiescence in mind; but at other times, as in the passage in the *Christian Virtuoso*, God's general concourse clearly meant an active sustenance. Boyle could not accept the idea that since the creation all things come to pass by the settled laws of nature alone. A law is a moral, not a physical, cause—a notional thing according to which an intelligent and free agent is bound to regulate its actions. "But inanimate bodies are utterly incapable of understanding what a law is, or what it enjoins, or when they act conformably or unconformably to it; and therefore the actions of inanimate bodies which cannot excite or moderate their own actions, are produced by real power, not by law, though the agents, if intelligent, may regulate

25. *Works*, 5, 520.

the exertions of their power by settled rules." [26] In other places Boyle mentioned the role of God in regulating the motions of the great heavenly bodies to keep them within their allotted orbits. Such words ill consorted with his favorite analogy of the clock. Perhaps the clock does not wholly express his conception of nature, and a figure that he drew in an early work may illuminate his thinking. "As the quill that a philosopher writes with," he suggested at that time, "being dipped in ink and then moved after such and such a manner upon white paper, all [of] which are corporeal things or their motions, may very well trace an excellent rational discourse; but [*sic*] the quill would never have been moved after the requisite manner upon the paper had not its motion been guided and regulated by the understanding of the writer." [27] The implication of these passages is clear: he was trying to maintain that God's continued participation in the operation of nature is necessary to the fulfillment of His natural laws.

Boyle tackled the major problem that the scientific theories of the 17th century posed for Christian beliefs: the reconciliation of providence and miracles with the mechanistic natural order. In its ultimate effect his solution of the quandary reduced providence to a benevolent original creation. The evolution of Boyle's thought, from the simple assertion that God can overrule the laws of His creation to the recognition that the Father contradicts Himself in doing so, reflected the changing opinion among the virtuosi as a whole, the progress from Charleton's position to the final arguments of Boyle. Following him, most of the virtuosi accepted and developed his final conclusions on the subject. Boyle also symbolized the attitude of the virtuosi in his reverent handling of the issue. His long effort to reconcile providence and miracles with natural law was only the introduction to further efforts by other virtuosi who were equally interested in the harmony of science and religion.

The great naturalist John Ray (1627–1705) developed an analysis of providence which ranked next to Boyle's in its insight and comprehension. Like Boyle, Ray had a clear perception of natural

26. Ibid., pp. 520–1.
27. Ibid., 2, 48.

order. His investigation of organic life unfolded a pattern in nature rising consistently from the lower orders of being to the highest forms. Ray was no more willing than Boyle to admit exceptions to the pattern; not the least important part of his work was his ruthless destruction of superstitions and tales of anomalies that seemed to violate it. Although his studies as a naturalist did not commit him necessarily to the mechanical philosophy, he adopted and expounded it in his book on natural religion, *The Wisdom of God Manifested in the Works of the Creation* (1691). Accepting the theory that physical reality is composed of tiny corpuscles in various conformations, he went on to picture the universe as a machine; and in his physiological essays he applied mechanical principles to the explanation of some organic functions. Nevertheless, because he looked through the eyes of a naturalist, his view of nature took on its own peculiar coloring which influenced his argument on providence.

Ray could not dismiss the apparent disparity between mechanical operations and organic life. Although he might use mechanics to explain aspects of physiology, he was too well acquainted with life to think that it could be reduced to mechanical terms; and he sought another factor in natural phenomena to account for it. In reply to the Cartesians he declared that animals are more than automata, and to illustrate the point he called upon his extensive store of natural observations. His posthumous *History of Insects* (1710) records an instance of a wasp burying a caterpillar which it had killed. After covering the hole, the wasp placed two pine needles as if to mark its location. "Who would not wonder in amazement at this?" Ray asked. "Who could ascribe work of this kind to a mere machine?" [28] He concluded that the more ordinary processes of life and growth also lie beyond the realm of physical mechanics. The pulse of the heart, for instance, is not a mechanical motion in Ray's opinion. It differs from voluntary motions in that the will does not control it. If the pulse is ascribed to the unconscious motion of the spirits, a cause of their reciprocal movement must be found; and in the end the explanation is forced back to some vital principle. A vital principle is also necessary to explain the vegetation of plants and such special cases

28. Cited in Raven, *Ray*, p. 396.

as the survival of slips cut from plants and thrust into the ground. To solve all of the problems, which was equivalent to explaining life, Ray had recourse to the principle of "plastic nature" worked out by Henry More and Ralph Cudworth, which postulated a spiritual vicegerent of God pervading the natural order and governing its operations.

Plastic nature became the foundation of Ray's discussion of providence. Not limited, as he conceived of it, to the role of a vital principle necessary for the existence of life in a mechanical world, plastic nature was made the support of the mechanical universe itself. Ray was troubled by the dualism of spirit and matter involved in the metaphysics of natural science; he tried to bridge the gap with the concept of plastic nature, restoring the primacy of spirit within the realm of physical nature. Blind inert matter seems incapable of performing even mechanical operations. How is matter to perceive the laws of motion and follow them? In his *Wisdom of God* Ray took exception to Boyle's idea of creation as given in the *Vulgar Notion of Nature*. Boyle said that God created matter and laws of motion, shaped matter into the parts of nature, and set the parts in motion.

> This hypothesis, I say, I cannot fully acquiesce in [Ray objected], because an intelligent being seems to me requisite to execute the laws of motion. For first, motion being a fluent thing, and one part of its duration being absolutely independent upon another, it does not follow that because anything moves this moment, it must necessarily continue to do so the next, unless it were actually possessed of its future motion, which is a contradiction; but it stands in as much need of an efficient to preserve and continue its motion as it did at first to produce it. Secondly, let matter be divided into the subtlest parts imaginable, and these be moved as swiftly as you will, it is but a senseless and stupid being still, and makes no nearer approach to sense, perception, or vital energy than it had before; and do but only stop the internal motion of its parts and reduce them to rest, the finest and most subtle body that is may become as gross and heavy and still as steel or stone. And as for any external laws or es-

> tablished rules of motion, the stupid matter is not capable of observing or taking any notice of them, but would be as sullen as the mountain was that Mahomet commanded to come down to him; neither can those laws execute themselves. Therefore there must, besides matter and law, be some efficient, and that either a quality or power inherent in the matter itself, which is hard to conceive, or some external intelligent agent, either God himself immediately or some plastic nature.[29]

Never willing to detract unjustly from another, Ray went on to quote the passage from the *Christian Virtuoso* in which Boyle also concluded that matter alone cannot obey laws, and Ray decided that they were in agreement. Plastic nature, Ray's own answer to the dilemma, is, then, the agent of God, not God Himself but the spiritual creation of God, delegated to implement natural laws. Pervading all of nature, the spiritual being both cherishes life in the bosom of matter, and controls the movements of material bodies. It vindicates the supremacy of spirit and unites all of creation into one organic whole.

While the concept of plastic nature appeared to solve the exclusion of God from His creation, it did little to redeem the doctrine of particular providence. Plastic nature was not the principle of fatherly care hypostatized into being. It was thought of as an impersonal force upholding the natural order, devoted to the rule of law, but not to the protection of individuals. It came to fulfill the law, not to destroy it. An idea of general providence was inherent in the concept, but in reinforcing the natural order it excluded the notion of particular providence. Ray never tried to reconcile particular providence with the determined natural order. As a firm believer in the Bible he accepted the miracles recorded there, but to him as to Boyle providence usually meant the benevolence of the original creation.

On occasion Ray could assert a more traditional idea of providence. His *Physico-theological Discourses* (1692) told of an earthquake in the West Indies, which he considered to be God's punishment of a sinful people. "For God does not stand by as an

29. Ray, *Wisdom of God,* pp. 54–5.

idle and unconcerned spectator and suffer things to run at random," he declared, "but His providence many times interposes and stops the usual course and current of natural causes. Nay, I believe and affirm that in all great and notable revolutions and mutations He has the greatest hand and interest, Himself ordering and governing them by His special superintendence and influence."[30] Ray never thought to resolve the obvious conflict between the statement in the *Discourses* and his scientific theories. In failing to do so, he repeated the attitude already remarked in Boyle, the preference to let inconsistencies remain if rigorous logic led to conclusions challenging religion. Although he endeavored to prove that the findings of science supported the truth of Christian doctrines, he did not feel obliged, when writing on religion, to check every statement against scientific theory. Thus he could continue to accept the doctrine of particular providence in works on religion while he spoke of an inexorable natural order in his writings on natural philosophy. As with Boyle, the final implication of his consideration of providence was to leave the Almighty as the upholder and preserver of a beneficent original order.

The mechanical conception of nature within which Boyle and Ray worked provided the framework of thought for nearly all of the virtuosi. As the scientific movement progressed, its hold became ever more strong. Some virtuosi such as John Wallis (1616–1703) and Robert Hooke (1635–1703) used it to suggest inquiries and to direct experiments without bothering to expound its theory. The success of their work in explaining natural phenomena did much to argue the validity of the mechanical hypothesis. Others made it the basis of speculations where factual evidence was not available. John Mayow (1640–79), for instance, attributed all natural phenomena to the mechanical operations of his favorite invention, the nitro-aerial particle; and the essays on vegetable physiology by Nehemiah Grew (1641–1712) attempted to explain organic processes by the mechanical movements of material particles. Whatever the scientific value of their specula-

30. John Ray, *Three Physico-theological Discourses* (3d ed. London, 1713), p. 271.

tions, which was small at best, they helped to spread the belief that natural phenomena arise from the size, shape, and motion of material bodies. All alike contributed to a deeper perception of nature's ordered course. Since the efforts to render particular providence compatible with the mechanical order were not successful, the difficulties of the doctrine of providence became only more acute as the scientific movement advanced. By the end of the century the virtuosi were finding it more and more difficult to admit any providential interruptions of nature, until finally some of them even denied the reality of miracles themselves.

The image of nature as an immutable machine led Nehemiah Grew, for instance, to ponder the doctrine of providence and the definition of miracles. "Providence," he decided, "is God's provision or forecast of causes sufficient to the fulfilling of all His ends." [31] God does not act in creation immediately by His direct power but works through the mediation of some one or more instruments, and it is senseless to suppose either that He made instruments which He does not use or that He failed to make enough instruments to perform His work.

> As the several parts of the universe are so many lesser engines, so the whole is not a mere aggregate or heap of parts, but one great engine having all its parts fitly set together and set to work, or one entire movement of divine art. To suppose it then either to stand still or to move irregularly in the whole or in any the least part without the supervening of a new divine power is to suppose the Author of it not to be His art's master. There is nothing therefore in Nature, neither miraculous nor anything else of the greatest moment, wherein God is to be thought a solitary and immediate agent. But that every thing depends upon some created cause or causes with commission or power sufficient to produce it. But as the causes which we see and contemplate, though of the most usual effects, do justly merit our adoration of the Supreme Cause, so more especially those which are unknown to us and whose effects are miraculous.[32]

31. Grew, *Cosmologia Sacra*, p. 85.
32. Ibid., p. 87.

There is no contradiction, Grew argued in another place, when philosophy teaches that something is done by nature which religion attributes to God. To say that the balance of a watch is moved by the next wheel is not to deny that the wheel and all of the watch's parts are moved by the spring or that the parts are made to move together by the one who designed them.

> So God may be truly the cause of this effect although a thousand other causes should be supposed to intervene, for all nature is as one great engine made by and held in His hand. And as it is the watchmaker's art that the hand moves regularly from hour to hour, although he put not his finger still to it, so it is the demonstration of divine wisdom that the parts of nature are so harmoniously contrived and set together as to conspire to all kinds of natural motions and effects without the extraordinary immediate influence of the Author of it. . . . But in that these things are not only made, but so made as to produce their natural effects, here is the sensible and illustrious evidence of His wisdom; and the more complicated and vastly numerous we allow the natural causes of things to be, the more duly we conceive of that wisdom which thus disposes of them all to those their effects. As the wisdom of the king is not seen by his interposing himself in every case, but in the contrivance of his laws and constitution of his ministers in such sort that it shall be as effectually determined of, as if he did so indeed. Thus all things are as ministers in the hands of God, conspiring together, a thousand ways towards a thousand effects and ends at one time, and that with the same certainty as if He did prepose that omnipotent fiat which He used at the creation of the world to every one of them.[33]

Grew's constant reversion to the mechanical simile is an interesting example of its domination over the minds of the virtuosi. The vision of the perfect machine so intoxicated him that he could not bring himself to admit any deviation from its regular opera-

33. Nehemiah Grew, *An Idea of a Phytological History Propounded* (London, 1673), pp. 101–3.

tion. He thought that every phenomenon could be traced to the machine.

This brought up the instance of miracles. Grew marched resolutely ahead where other virtuosi had balked; drawing the logical conclusion from his reasoning, he denied that the biblical miracles were supernatural acts. A miracle—he continued to use the word—is defined by four conditions. First, its cause must be unknown; and secondly, the effect itself must be extraordinary. Thus an eclipse is not miraculous, since we know the cause; nor is the attractive power of a magnet, which we may observe every day. An unknown cause and an extraordinary effect do not add up to the supernatural, however, and Grew scanned the list of biblical miracles, pointing out from his scientific knowledge the natural causes that had probably been in action. The Nile flowed with blood because a disease attacked the fish. The walls of Jericho fell under the jolt of an earthquake. To the people at the time both events had appeared to be supernatural; to the scientist of the 17th century they stood revealed as the natural products of natural forces. The third condition of a miracle restored some of the sullied luster: a miracle must occur in its proper setting, limited as to person, place, and time. Only Moses could call on the Nile to run blood; the Egyptian magicians could not match his works. Not every river, only the Nile, turned to blood; and in time the event coincided with a host of other miracles and with the exodus of Israel from Egypt. The circumstances, Grew concluded, manifest the ultimate hand of God. Finally, miracles must be directed to a suitable end, as the biblical miracles testified to God's power and protection.[34] Grew's explanation of the miraculous reduced it to the level of another material event, an effect built into the machine at the original creation, a product of God's hand to be sure but no more so than the rest of nature. "Nature itself," he declared, "is a standing miracle, the operation whereof we should as much wonder at as any miracle if we did not see them every day. . . . But as the ordinary effects of nature are miracles in course, so miracles, specially so called, are natural effects either out of course or otherwise extraordinary, *viz.* by the ordination

34. Grew, *Cosmologia Sacra,* pp. 194–8.

and authority of the supreme cause. Wherefore, as the asserting of miracles does not derogate from the wisdom of the creation nor the majesty of the Creator, so neither does the intervening of natural causes overthrow miracles." [35]

Another of the virtuosi, Edmond Halley (1656–1742), who belonged to a generation still later than Grew and who continued his activity well into the 18th century, arrived at a similar conclusion in regard to miracles. A great astronomer and investigator of navigational aids, as well as a leader in other fields of science, Halley was one of the brightest lights among the virtuosi. He was not, however, a devoutly religious man. During his life the rumor spread that he was a skeptic or even possibly an atheist. His reputed free thinking led the University of Oxford to reject his application for the Savilian chair of astronomy in 1691, and one Scotsman reportedly traveled all the way to London to see the man with less religion than David Gregory, to whom the Savilian appointment went. From the scanty evidence available the charge of atheism appears to have been unfounded; but those who accepted it may have found some justification, at least for the charge of skepticism, in Halley's essay on the deluge. Delivered before the Royal Society in 1694, the essay was printed much later in the *Philosophical Transactions*. Of necessity it embodied an attitude toward the miraculous. Since Halley left no treatises on religion —in itself a possible indication of cold-heartedness in his day and age—and no full statement of what he believed in regard to miracles, the paper on the deluge is the primary evidence of his opinion. Fortunately, he made his position clear, and no fuller statement is needed. Halley's essay was a reply to a suggestion by Robert Hooke that the deluge had been caused by a compression of the earth into a prolate spheroid which forced the water out of the interior. From a scientific point of view Halley found the explanation inadequate; only the two poles would be covered with water if such a compression of the globe occurred. Furthermore, he added—and here he began to impinge on religious beliefs—"such a supposition cannot well be accounted for from physical causes, but requires a preternatural *digitus Dei*, both to compress, and afterwards to restore the figure to the

35. Ibid., p. 316.

globe." Since the Almighty generally makes use of "natural means to bring about His will," Halley continued, some natural force must have caused the deluge.[36] John Ray assigned natural causes to the deluge, but he manifestly considered the flood a basically supernatural event. Halley's attitude on the other hand pointed to a full rejection of the miraculous frame of mind. That the deluge actually occurred he did not question. That it was anything but a natural phenomenon he could not believe. In Halley's conception the compromise with the biblical miracles was ended and the last vestige of particular providence in the material realm disappeared.

From the vantage point of the 20th century it appears that Halley's position was the inevitable conclusion toward which the virtuosi moved. As they studied nature and found the natural causes of natural effects, they restricted the area of the supernatural more and more. For those events to which as yet they could assign no cause they assumed that some natural cause did exist. The compromise which tried to assert the inexorable order of nature and to maintain the reality of miracles in biblical times was bound to fail. If miraculous intervention in the natural order in the 17th century was incredible, miraculous intervention two thousand years earlier was no less incredible. Nature was shown to perform her own work according to her own laws, and natural science, in revealing those laws, destroyed any argument for supernatural causation. Floods, earthquakes, disease—all of the extraordinary events that the untutored imagination might have attributed to divine causation—were seen to be as natural as the daily rotation of the earth. The decision of Halley and Grew to deny the supernatural element in even the biblical miracles was only the logical conclusion from the principles shared by every virtuoso.

The question arises whether the virtuosi did any injury to religion. Insofar as the scientific movement of the 17th century challenged those who found the hand of God in every unusual event and insofar as it cut away the tumorous growth of super-

36. Edmond Halley, "Some Considerations about the Cause of the Universal Deluge," *Philosophical Transactions*, 33, 118; reprinted in *Philosophical Transactions Abridged* (11 vols. London, 1731–56), 6, Pt. II, p. 3.

stition, it performed a salutary operation on religion. The essentials of Christianity did not lie in the belief that God was an arbitrary agent in the world of nature directly producing all of the multiple events that make up the world's existence. In helping to complete the picture of an ordered creation worthy of an omniscient Creator, the virtuosi did not destroy any foundation stone of the Christian temple.

Of course, the traditional meaning of the doctrine of providence transcended the level of popular superstition. The virtuosi recognized providence as an essential Christian doctrine; as practicing Christians they would not deny it. To preserve it they took the only route that their conception of nature left open, interpreting providence as the general benevolence of the original creation and the continued sustenance of the natural order. The current of particular providence was swallowed up in the ocean of general providence. Nevertheless the virtuosi did insist on retaining the idea of divine providence. Their work was to reinterpret it in the light of new knowledge, not to destroy it and cast it out. Again they were performing an essential service for Christianity, in effect modernizing it, keeping it abreast of the progress of learning, refusing to let it become incredible to the educated, intelligent man. As long as the view is restricted to the physical world, their ultimate interpretation of divine providence appears to have been the only one compatible with their scientific theories.

The nub of the matter did not lie in the peculiar mechanical conception of nature held by most of the virtuosi; it lay rather in the fact that every profound investigation of natural phenomena has ultimately concluded that there is an order in nature unbroken at least within the range of its experience. Thomas Sydenham (1624–89) arrived at the same conclusion even though he rejected the mechanical conception of nature. As an Aristotelian in philosophy he recognized laws governing the development and action of created beings, and in his essay on natural religion he addressed himself to the question of providence. It would be unreasonable, he thought, to expect God to upset the physical creation in order to satisfy the desires of one person; for instance, He will not restore the vigor of youth to an aged body or do "such other things which in the order of nature are put under a necessity of not being able to come to pass." In events that do not contradict the natural order, however, God retains His freedom of action. Should I be shipwrecked far at sea, I should undoubtedly be drowned, "yet towards the preserving me from this mischief He may be pleased so to dispose the previous circumstances of my will and other things, as to prevent my going to sea, and so in this and in other things He may hinder the occasions leading to my destruction." As far as irrational beings are concerned, the Creator has determined them to act in some uniform course suitable to themselves and to the whole of nature; but there is nothing more derogatory to God than to conclude that He is not a free agent. "And though He has set up certain lights in intellectual natures, yet having given these a liberty of will incident to the very nature of reasonable beings, He retains His power of inclining or not inclining such intellectual natures to pursue courses leading to their welfare. And truly, in the nature of intellectual creatures seems to be the dominion of the Divine Agent, which though He does not determine in their operations as He does inferior beings, yet He does, when He pleases, make their lights set up in their minds more illustrious, or else by a peculiar incitement of thoughts render them more disposed to comply with their own good and to avoid their own unhappiness, so laying in the mean time all the train of circumstances without them as that they may conduce to this end." [37]

Sydenham both agreed with the other virtuosi and indicated an important consideration that did not appear in their discussions. While he concurred in their conclusion that God will not upset the laws of the material world, he went on to assert that divine providence can still operate within the human soul. Determination in the physical world does not imply determination in the spiritual realm. There is no more glaring omission in the virtuosi's protracted consideration of providence than their failure to seize on this point which was fully compatible with the Cartesian dualism accepted by them. They affirmed the existence of the non-material, both in God and in man, yet they discussed providence only in regard to the material world. Although they

37. Sydenham, *Works, 2,* 310–11.

pondered the doctrine of providence honestly and even reverently, their consideration of it was curiously one-sided, as though the major problem of Christianity were the relation of God to physical nature. To be sure, it was a problem, and the virtuosi were forced by their studies to consider it, but by limiting themselves to that issue they were unable to preserve the idea of particular providence.

Although the disparity between the doctrine of particular providence and the theory of a natural order of events was not the peculiar problem of the mechanical conception of nature, the virtuosi's attempts to resolve the difficulties of God's relation to the physical world did arise in part from the mechanical hypothesis. Inherent in the mechanical conception of nature was a deeper challenge to Christianity and to all religion. The Cartesian metaphysics upon which it rested, the duality and independence of spirit and matter, involved the elevation of the physical world to an autonomous order. The spiritual control of the creation had to be asserted if the drift toward materialism was to be checked. Two of the virtuosi addressed themselves directly to the danger of materialism, though their efforts were less successful than their labors in reconciling providence with the natural order. Boyle affirmed that the cosmic machine is not autonomous because blind matter by itself cannot obey laws. John Ray conceived of a plastic spirit informing the movements of senseless matter. But these were mere afterthoughts imposed on a picture already completed. They dreamed of a universal machine perfect in every part and functioning, by their own admission, without the attendance of mechanics or operators; and then they tried to insert the presence of spiritual control without disturbing the mechanical wonder. The compelling picture of a perfect natural machine, given the presumption of validity by the remarkable discoveries of the 17th-century scientists, was not inextricably bound up with the philosophical assertions of Boyle and Ray that tried to save the primacy of spirit. Minds less devout might simply strip away the veneer of spiritual control and declare that the machine is wholly material, an autonomous mechanism. Pure materialism was not an impossible deduction from their work, and the 18th century produced men like Baron d'Holbach who insisted on making it.

The virtuosi themselves would not and did not make that deduction. If men of the 18th century looked upon the world as an autonomous machine, the Christian virtuosi affirmed that divine sustenance is necessary to its operation. Their final idea of general providence, painfully worked out by a number of thinkers, was more than a rejection of the doctrine of particular providence. In asserting the primacy of spiritual control, it was a true reconciliation between science and religion.

CHAPTER 5

The Growth of Natural Religion

> The belief of a Deity being the foundation of all religion . . . it is a matter of the highest concernment to be firmly settled and established in a full persuasion of this main point; now this must be demonstrated by arguments drawn from the light of nature and works of the creation.
>
> Ray, *The Wisdom of God Manifested in the Works of the Creation*

> It is granted on all hands that our chief end, our *summum bonum,* is our own happiness . . . So that according to this principle, to aim chiefly at our own well-being is not only permissively lawful . . . but 'tis likewise essentially necessary to our very nature . . .
>
> John Wilkins, *Sermons Preached upon Several Occasions*

IF THE VIRTUOSI realized that their concept of natural order seemed to conflict with the doctrine of providence and required some form of reconciliation with it, they did not recognize a graver threat to Christianity in the very means by which they sought to defend it. Faced with a supposed growth of atheism, they seized the opportunity to apply their own precept that natural philosophy is an aid to religion. By demonstrating the existence of God with unanswerable proofs, they thought that they would utterly destroy the menace of atheism. Natural religion was supposed to be the sure defender. Yet in the end the defender turned out to be the enemy within the gates. In theory natural religion was meant to supplement Christianity, to provide it with a rational foundation; in practice it tended to displace it. While the virtuosi concentrated vigorously on the demonstrations of natural religion and proved to their own satisfaction that the cosmos reveals its Creator, they came to neglect their own contention that natural religion is only the foundation. The supernatural teachings of Christianity received little more than a perfunctory nod, expressing approval but indicating disinterest. Although the absorption in natural religion and the external manifestations of divine power did not dispute or deny any specific Christian doctrine, it did more to undermine Christianity than any conclusion of natural science.

The primary impulse toward natural religion among the virtuosi came from the side of Christianity itself and not from covert skepticism or rationalism. Disgust with sectarian excesses, expressing itself in religious rationalism, was not absent, but in most of the virtuosi it was reinforced and overshadowed by a positive concern to preserve religion from the danger of atheism. In this they reflected the concern that had led to the development of natural religion among the Scholastics, although the virtuosi's adversary, "atheism," differed of course from the paganism against which the Scholastics had contended. On the surface it appears strange to find a vivid apprehension of atheism in an age not noted for its disbelief. That there was apprehension is clear from the books that the virtuosi wrote. Both Walter Charleton and John Wilkins published refutations of atheism. Several of Boyle's works addressed themselves to the threat, and scarcely a single one of his multitudinous writings did not touch on the subject. John Ray, Nehemiah Grew, and Joseph Glanvill all joined in the disputation.[1] Most of the other virtuosi controverted infidelity in brief digressions if they did not devote whole works to the problem. England, Walter Charleton wrote in 1652, "has of late produced, and does at this unhappy day foster, more swarms of atheistical monsters . . . than any age, than any nation has been

1. Charleton, *The Darkness of Atheism Dispelled by the Light of Nature; The Immortality of the Human Soul, Demonstrated by the Light of Nature.* Wilkins, *Of the Principles and Duties of Natural Religion.* Boyle, *The Usefulness of Experimental Philosophy,* Oxford, 1663 (primarily a discussion of its usefulness for religion); *A Discourse of Things above Reason,* London, 1681; *A Disquisition about the Final Causes of Natural Things,* London, 1688; together with innumerable passages in other works and a large number of unpublished papers. Ray, *The Wisdom of God Manifested in the Works of the Creation; Three Physico-theological Discourses,* under the the original title of *Miscellaneous Discourses Concerning the Dissolution and Changes of the World;* and *A Persuasive to a Holy Life,* London, 1700. Grew, *Cosmologia Sacra: or a Discourse of the Universe as It Is the Creature and Kingdom of God.* Glanvill, *Saducismus Triumphatus: or Full and Plain Evidence Concerning Witches and Apparitions,* London, 1666, under the original title of *Philosophical Considerations concerning Witchcraft;* and *Philosophia Pia, or a Discourse of the Religious Temper and Tendencies of the Experimental Philosophy of the Royal Society,* London, 1671.

infested withal." [2] "There is no one that is not very much a stranger to the world," Glanvill repeated nearly thirty years later, "but knows how atheism and infidelity have advanced in our days and how openly they now dare to show themselves in asserting and disputing their vile case." [3] The works mentioned here came from the pens of the virtuosi alone; they could be supplemented by a larger number from men outside the group. One book, or even two, might be blamed on the personal idiosyncrasies of the authors, but so large a number implies some general cause.

The virtuosi themselves made the source of their anxiety clear. Atheists were materialists, Epicureans, those who held that the universe is the chance concourse of atoms in motion. Atomism had been the philosophy of atheism, or at least what the 17th century considered to be atheism, in the hands of Epicurus; now after a long sleep atomism was awakening in Western Europe, and the threat of atheism sprang from the same bed. It appears remarkable at first to find the virtuosi attacking Epicureanism, since they themselves were the leading advocates of Epicurus' atomism. It does not seem strange when the tone of personal concern, almost of bad conscience, is noticed in their refutations of atheism. On philosophical grounds they were convinced atomists. To give up atomism would in their view have been to give up any serious pretension of studying natural science, but at the same time they recognized that they were playing with fire in championing a philosophy that had once stood in opposition to their fundamental religious beliefs. Hence their endlessly repeated confutation of materialistic atheism takes on the appearance of a quest for personal certainty, of a reply to the small voice of doubt that had crept into their minds with the atoms. They were proving to themselves, as well as to society at large with its articulate critics, that atomism did not lead to atheism, that it had been reformed in their hands, that it was now in the service of the Christianity which they, as well as their critics, believed in and revered.

As if divine providence were intent on pricking their consciences, the virtuosi saw their worst fears embodied before their

2. Charleton, *Darkness of Atheism,* "To the Reader."

3. Joseph Glanvill, *Saducismus Triumphatus* (London, 1681), Pt. II, p. 1.

eyes. The materialistic philosophy of Thomas Hobbes was everything that they did not want atomism to be. In Hobbes they saw contemporary proof of the perversion to which atomism was susceptible, and the horrendous figure of the sage of Malmesbury lurks behind the references to atheism that season the works of the virtuosi. The excitement caused by the appearance of Hobbes' philosophy in the middle of the 17th century was only dying down when the 18th century dawned. The refutation of Hobbes became the standard performance in Christian gymnastics until Charles II likened him to a bear against whom the Church played her young dogs to give them exercise. And not the Church alone, for the virtuosi were perhaps even more concerned to destroy a system so nearly resembling their own. Had Hobbes never existed, the virtuosi would probably have felt impelled to prove that atomism is not equivalent to atheism. With Hobbes the imperative was doubled; he was the concrete realization of abstract fears. John Wallis expressed the concern of the virtuosi when he wrote to Huygens in 1659 explaining his virulent battle of pamphlets with Hobbes. "Our Leviathan is furiously attacking and destroying our universities (and not only ours but all) and especially ministers and the clergy and all religion," Wallis reported, "as though the Christian world had no sound knowledge . . . and as though men could not understand religion if they did not understand philosophy, nor philosophy unless they knew mathematics. Hence it seemed necessary that some mathematician should show him by the reverse process of reasoning how little he understands of the mathematics from which he takes his courage."[4] Glanvill considered that "Atheists, Sadducees . . . [and] Hobbists" were all of the same species, meaning by Sadducees those who deny the existence of spirits.[5] Writing at a later date, Nehemiah Grew placed Spinoza beside Hobbes; and the virtuosi also feared the religious implications of Descartes' system. The Frenchman falls into a special category, however, for the virtuosi admired him greatly as a philosopher and hence were

4. Wallis to Huygens, January 1, 1659; cited in J. F. Scott, *The Mathematical Work of John Wallis, D.D., F.R.S. (1616–1703)* (London, Taylor and Francis, 1938), pp. 170–1.

5. Glanvill, *Saducismus*, Pt. II, p. 8.

not prone to call him an atheist. His conception of nature as a machine was of course basic to their scientific thought. What the virtuosi objected to in Descartes was his theory that the world had formed itself through the movement of material corpuscles guided by laws of motion, whereas they insisted that God had both created matter and molded it into the framework of nature. In their demonstrations of natural religion they constantly elaborated this point, proving that not only matter and natural laws of motion alone but individual creatures as well were the work of God. For the most part, however, the virtuosi were content to name their adversary by implication; and when they argued against the theory that the world was the chance production of matter in motion, they usually attacked both the general possibility of atomistic materialism and the specific reality of Thomas Hobbes.

Added to the dangers of the Epicurean philosophy and reinforcing their fear of it was the practical Epicureanism of Restoration England. When atheism was mentioned by the virtuosi, moral laxity was nearly always coupled with it.

> Some of the learned heathen [John Wilkins wrote in *Natural Religion*] have placed the happiness of man in the external sensual delights of this world; I mean the Epicureans, who though in other respects they were persons of many excellent and sublime speculations, yet because of their gross error in this kind they have been in all ages looked upon with a kind of execration and abhorrency, not only among the vulgar but likewise among the learneder sort of philosophers. 'Tis an opinion this, so very gross and ignoble, as cannot be sufficiently despised. It abates the understanding of man and all the principles in him that are sublime and generous, extinguishing the very seeds of honour and piety and virtue, affording no room for actions or endeavors that are truly great and noble, being altogether unworthy of this nature of man, and reduces us to the condition of beasts.[6]

Wilkins found this aspect of Epicureanism only too prevalent around him, and he concluded that the true reason moving those who reject Christianity "is the strictness and purity of this re-

6. Wilkins, *Natural Religion*, pp. 403–4.

ligion, which they find puts too great a restraint and check upon their exorbitant lusts and passions." [7] Among the papers of Robert Boyle are a number of fragments on the causes of atheism, attributing disbelief to the vicious and sinful life that the infidels lead.[8] The vileness of some men's lives, Boyle decided, makes it difficult for them to give up skeptical opinions, since to do so they must not only confess that they have been mistaken and wicked in the past "but must for the future forsake those practices which they most doted on and wherein they placed their happiness." [9] John Wallis spoke of "debauched persons (with whom atheism and ribaldry pass for wit)," and in a letter to Bishop Tenison he mentioned Rochester, the most notorious rake in the court of Charles II, as a leading atheist.[10] William Petty called scoffers "libertine skeptics." [11] Religious men have always found declining morals in the society around them, but few have had more cause for alarm than the critics of Restoration England. Moral degeneration furnished practical evidence of the spread of Epicureanism, and it served to heighten their concern that Epicurean atomism might usher in the triumph of atheism.

To refute the claims of atheism the virtuosi adopted the procedure outlined by Wallis in his letter to Huygens, employing the materials of natural philosophy to prove that the basis of atheism was rotten and unsound. In a version of his essay *Philosophia Pia* Joseph Glanvill contended that the pretensions of the modern atheist could be refuted only by those who understood the atheist's system of ideas. For the champions of Christianity to rely on forms, elements, occult qualities, and heavenly influences to overwhelm mechanical materialism would be fatuous and vain.[12] Robert Boyle and John Ray, among others, put his advice to practical use. When Boyle adopted atomism in his first philo-

7. Ibid., p. 408.

8. Robert Boyle, "Fragments on the Causes of Atheism," in Royal Society, Boyle Papers, 2, fols. 78–9, 147–50.

9. Robert Boyle, "Four Conferences about as Many Grand Objections against the Christian Faith," in Royal Society, Boyle Papers, *1*, fol. 45.

10. John Wallis, *The Resurrection Asserted: in a Sermon Preached to the University of Oxford on Easter Day, 1679* (Oxford, 1679), p. 29. Wallis to Bishop Tenison, November 30, 1680; British Museum, Birch MS 4, 292, fols. 2–5.

11. Petty to Southwell, October 30, 1676; *The Petty-Southwell Correspondence, 1676–1687*, ed. Marquis of Lansdowne (London, Constable, 1928), pp. 6–7.

12. Glanvill, *Essays*, Essay 4, pp. 7–8.

sophical work, *The Usefulness of Experimental Philosophy,* written between 1649 and 1653, he addressed himself to the atheistical conclusions of Epicurus and Lucretius. The two ancient philosophers had accepted a number of premises which Boyle did not find justified—that matter is eternal; that from eternity it has been divided into atoms; that the number of atoms is infinite; that they have infinite space to move in; that they are endowed with a great variety of shapes; that they have been in motion from eternity, falling and inclining toward each other; and that given all this, nothing but fortuitous concourse is needed to produce all of the bodies that compose the world. With justice Boyle claimed that Epicurus and Lucretius had assumed everything that had to be proved. It was God Who made the atoms, put them in motion, and formed them into the creatures of nature. "And really," Boyle concluded, "it is much more unlikely that so many admirable creatures, that constitute this one exquisite and stupendous fabric of the world, should be made by the casual confluence of falling atoms jostling and knocking one another in the immense vacuity, than that in a printer's working-house a multitude of small letters, being thrown upon the ground, should fall disposed in such an order as clearly to exhibit the history of the creation of the world . . ." [13] In a manuscript paper "On the Conversion of Atheists" he argued further that the functions of mind can never be reduced to the mere concourse of material atoms.[14] John Ray took up the matter where Boyle left it, building his case against atheism with evidence drawn from organic life. The complexity and harmony of the parts of plants and animals and their constancy of reproduction imply a deeper cause than the chance meeting of blind atoms. The formation of a foetus "is so admirable and unaccountable that neither the atheists nor mechanic philosophers have attempted to declare the manner and process of it, but have . . . very cautiously and prudently broken off their systems of natural philosophy here and left this point untouched." [15] The whole case against pure

13. Boyle, *Works,* 2, 43.

14. Robert Boyle, "On the Conversion of Atheists," in Royal Society, Boyle Papers, 2, fols. 18–19.

15. Ray, *Wisdom of God* (4th ed.), pp. 340–1.

materialism was summed up in one of Dr. Charleton's few pithy phrases. If I knew an atheist, he asserted, "I would do my best to bring him into this theater [for anatomical dissection], here to be sensibly convinced of his madness." [16] In one form or another most of the virtuosi echoed these words. The organization of nature was proof that she had not been the unplanned work of atoms in motion, and natural philosophy itself could take the lead in refuting those who tried to make it serve atheism.

The compulsion of self-defense also helped to inspire the virtuosi's campaign to destroy disbelief. In spite of their protestations of orthodoxy some of the virtuosi were tarred with the brush of atheism; they were not anxious to carry that mark in 17th-century England. The charge evidently struck Walter Charleton. A preface entitled "An Apology for Epicurus," attached to his translation of *Epicurus's Morals* (1656), limped under a load of self-justifying humility. A year earlier in 1655 Charleton had published anonymously an uninspired story called *The Ephesian Matron,* which had served as a vehicle for expounding Epicurean ethics. Perhaps his authorship had become known—although it is difficult to imagine such an obscure book feeding much of a fire —or perhaps his advocacy of the Epicurean philosophy and his known friendship with Thomas Hobbes had besmirched his name. Whatever the cause, he was obviously concerned to dissociate himself from those doctrines of Epicurus which he, as he loudly protested, considered as reprehensible as the next man. Joseph Glanvill's *Plus Ultra* (1668) was written under a similar cloud of suspicion. An earlier work, the famous *Vanity of Dogmatizing* (1661), had brought down wrath upon his head; he had been charged with atheism. In *Plus Ultra* he defended both natural philosophy and his own interest in it.

To such accusations of infidelity the virtuosi were tender in the extreme. Edmond Halley learned how damaging they could be when his candidacy for the Savilian chair in astronomy in 1691 ran aground on his reputation of disbelief. Being somewhat impetuous, Halley was rash enough three years later to read his

16. Walter Charleton, *Enquiries into Human Nature in Six Anatomic Prelections in the New Theatre of the Royal College of Physicians in London* (London, 1680), preface.

liberal opinion on the deluge to the Royal Society. When he saw that his paper was provoking controversy, however, he beat a hasty and humiliating retreat which suggested that he did not care to risk another clash with aroused orthodoxy. In the original paper Halley maintained that the biblical account of the deluge was incomplete because the causes given could not possibly have produced the flood. Halley himself suggested that a comet had hit the earth, changing the axis of rotation and causing the seas to move their position. At the next weekly meeting of the Royal Society he rose to add some "further thoughts." On the advice of "a person whose judgment I have great cause to respect" he wished to say that his opinion was not meant to apply to the biblical deluge at all. He had merely been speculating about the effects of such an accident as might have befallen the earth at an earlier time and might have reduced a former world to chaos from which this world was formed. It was a strange conclusion to a paper which had started out to explain the natural causes of the deluge. The second paper would probably not have pleased the orthodox much more than the first, although it did avoid biblical criticism. Halley was wise to obey his fear of clerical censure and to withhold both papers from publication for thirty years, until they finally appeared in the *Philosophical Transactions* in 1724.[17]

William Petty learned his lesson from the example of others. For some time he worked on a treatise, *The Scale of Creatures,* which dealt with some points of natural religion. The more he wrote, the more he worried about its reception. "I have of late," he confided to his friend Robert Southwell, "learned some discretion of persons, times, and places, and that truth and sincerity signify little. Wherefore if I write *in re ardua* anything heterodox, what shall I get if I do well?—Rx 0000. What shall I suffer if it please not the censorious and envious, the sciolous, &c.?—Rx 99999999 etc."[18] Petty eventually chose not to publish at all rather than lay himself open to criticism.

The timidity evident in the cases of Halley and Petty sprang from a concern that all of the virtuosi must have known, a fear

17. *Philosophical Transactions,* 33 (1724); reprinted in *Philosophical Transactions Abridged,* 6, Pt. II, pp. 1–4.

18. Petty to Southwell, August 4, 1677; *Petty-Southwell Correspondence,* p. 31.

that they should be tagged with the label of religious skeptic. Petty decided not to publish; and since his treatise has not survived, it is impossible to know whether there was anything exceptional in it. But silence was not the usual defense. If they were going to pursue their scientific investigations, they had to meet the suspicions of an intolerant age. One of the reasons behind the virtuosi's ventures into natural religion to refute atheism was the desire to preserve their own reputations for orthodoxy.

A complex of closely associated motives impelled this group to investigate natural religion in order to defend the fundamentals of Christianity. A breeze from a different quarter—the excesses of religious zeal that 17th-century England also experienced—helped to drive them into the same harbor. Religious strife that sundered English society in the first half of the century and culminated in the Civil War set many devout men on the search for fundamentals on which all could agree, and the extravagance of the sects prompted them to seek a firm rational basis for their faith. The growth of rational theology was not primarily a movement of the virtuosi, but they were influenced by the same considerations and participated in the movement. Perhaps their taste for rational investigation increased their appetite for rational religion above the average. Young Robert Boyle was shocked by the sectarian fury in London when he returned to England from the continent at the height of the Civil War. "If any man have lost his religion," he wrote to his friend Francis Tallents, "let him repair to London, and I'll warrant him, he shall find it; I had almost said too, and if any man has a religion, let him but come hither now, and he shall go near to lose it . . ." [19] To John Durie he lamented "that men should rather be quarreling for a few trifling opinions, wherein they dissent, than to embrace one another for those many fundamental truths, wherein they agree." [20] Although Boyle's main compulsion toward natural religion was the desire to establish defenses against disbelief, he fully appreciated the need for rational foundations secure from the vagaries of enthusiasm.

Joseph Glanvill vented his disgust with sectarian fanatics in

19. Boyle to Tallents, February 20, 1647; Boyle, *Works, 1,* xxxv.
20. Boyle to Durie, 1647; ibid., *1,* xxxix–xl.

the collection of his essays, which he published in 1676. The concluding discourse, "Anti-fanatical Religion and Free Philosophy," purported to be a continuation of Bacon's *New Atlantis;* and in picturing utopia Glanvill presented an allegory of England's religious upheavals. In the land of Bensalem a sect called the Ataxites arose claiming greater purity than others and direct access to God. Gradually they weaned the people away from the established religion, until finally they provoked a revolution in which they executed the king. In the ensuing confusion all sorts of queer sects appeared and did battle with each other, each asserting its special revelation and communion with God. The confusion also fathered a more substantial group, the Young Academicians, the heroes of the fable. The Young Academicians drew one advantage from their unhappy times: "they were stirred up by the general fermentation that was then in men's thoughts and the vast variety that was in their opinions to a great activity in the search of sober principles and rules of life." Believing in free inquiry and hating dogmatism, they studied the modern thinkers and natural philosophy as well as the ancient church fathers. Against the sectarians they asserted the rights of reason as necessary to uphold religion, and against the atheists and infidels they used reason to prove the essentials. The Young Academicians believed that the fundamentals of religion were the most important doctrines, and they refused to be led astray after inconsequential minutiae. Religion, they argued, was not meant to furnish matter for disputation; it was intended to perfect men's lives. The sectarians had bound themselves so tightly within their own doctrines that they had forgotten charity. Religion was not perfecting their lives.[21] The Young Academicians in Glanvill's fable represented the Cambridge Platonists, but the story illustrates the reaction of a wider group to sectarian madness.

The exploration of natural religion harmonized so readily with the predilections of the virtuosi that they would probably have taken it up had there been no pressures on them. They were only too ready to see God in the works of nature. From the pious exclamations at the wonders of creation, with which they punctuated their writings, it was only a step to the fully elaborated

21. Glanvill, *Essays,* Essay 7.

demonstration of natural religion. Indeed natural religion was only the converse of their proposition that science and religion do not conflict with each other. Nature is God's second revelation, they said. The Almighty has revealed His will in the Scripture; He has revealed His wisdom and power in nature. What could have been more obvious than the conclusion that religion could be founded on the natural revelation as well as on the scriptural one? And what could have been more characteristic and instinctive in a group which discovered the hand of God in every production of nature? Prone as the virtuosi were to see phenomena as natural miracles, awe-struck by the wonders of nature, led to spontaneous worship by the divine wisdom revealed in creation, they could hardly have refrained from transforming their studies into religious truths. The virtuosi's employment of natural religion in defense of Christianity was not in itself anything new; the alliance of Christianity with natural religion was centuries old. The distinctive feature of natural religion in their hands was the prestige and importance that they assigned to it; and their inclination to worship God through the creation, together with their growing confidence in the power of human reason, is of major significance in explaining the importance that they attached to natural religion. The intellectual atmosphere of England during the late 17th century reinforced the consequence of natural religion. Against a background of sectarian controversy which called into question the possibility of finding absolute authority in the Bible, and of enthusiastic vagaries which challenged the faith of reasonable men, the natural revelation promised the means of restoring certainty.

Many reasons, then, operated to direct the attention of the virtuosi toward natural religion, and religious skepticism was not among them. They thought of natural religion as a positive aid to Christianity, securing its foundations and perhaps purging it of superstition but not disproving or displacing it. In concentrating on the rational and demonstrable fundamentals of religion, however, they tended to forget in fact what they admitted in principle —that natural religion is only the foundation for the doctrines of Christianity that transcend rational proof. Christianity as it left their hands was a reasonable religion for reasonable men with the spiritual profundity of traditional Christianity thrust to the side.

The first virtuoso to discuss natural religion thoroughly was Walter Charleton, who covered all of its aspects in a series of books. In the preface to the first, *The Darkness of Atheism Dispelled by the Light of Nature* (1652), he condemned those "who think any principle of religion either profanely disparaged and debased, or implicitly convelled or staggered when brought to the test of natural reason, though only for confirmation." [22] He freely admitted that there are mysteries in Christianity transcending reason (although he neglected them in practice), and he agreed that the Bible is the divinely inspired source of Christian truth. Nevertheless his head rested a little more softly on the pillow at night when he knew that reason confirms what the Scriptures affirm. Following the spirit of his preface, Charleton tried to bring the support of reason to Christian principles. *The Darkness of Atheism* demonstrated the existence of God and the reign of providence. *The Immortality of the Human Soul, Demonstrated by the Light of Nature* (1657) established the soul's immortality by rational proof. Three separate books surveyed morality—his translation *Epicurus's Morals* (1656), which was really an attempt to reconcile Epicurean ethics with Christianity; *The Natural History of the Passions* (1674); and *The Harmony of Natural and Positive Divine Laws* (1682).

Of necessity natural religion could not deal with the central doctrine of Christianity, the mediation of Jesus to redeem man. Charleton's discussion emphasized the omission. In stressing the creative and gubernatorial aspects of God, he evoked the image of a powerful figure standing over, ruling, and judging man, but aloof from communion with him. Charleton's God was not the loving Father Who had sent His Son as a sacrifice for man's sins. The doctor's treatment of rational ethics tended in the same direction. Under the inspiration of the ancient moral philosophers he adopted an ethic of rationally guided self-interest, which bore no trace of Christian love and which set a goal to be attained by natural means alone.

Charleton's ethics were adumbrated in a discussion of providence and free will in *The Darkness of Atheism.* He held that

22. Charleton, *Darkness of Atheism,* "To the Reader."

free will is the necessary companion of rationality, and that liberty of election is really the property of the intellect, which guides the will. The will invariably pursues the good that the intellect chooses. The imperfection of human life lies in the fact that the intellect is often deceived by evil masquerading as good and leads the will from the paths of virtue. God, Who recognizes human weakness, has promulgated rewards and punishments to direct the intellect toward virtue. In the *Darkness of Atheism,* Charleton did not develop the hedonism that was latent in his discussion, but in his principal work on moral philosophy, his volume *Epicurus's Morals,* it dominated the discussion. *Epicurus's Morals* presented the familiar arguments of hedonism. Pleasure is good, pain evil. The supreme goal of man is felicity, the state in which he enjoys the maximum of pleasure with the minimum of pain. In itself every pleasure is good; even a life of gross corporeal pleasures would be felicity if the pains connected with them could be avoided. Since the pains cannot be avoided, human happiness is found, not in active titillation of the senses, but in constant calmness, placability, and immunity from distresses, a life free from bodily pain and mental perturbation. Happiness is attained through the life of virtue and the dominance of reason. While *Epicurus's Morals* defined the state of happiness to which virtue leads, Charleton offered some advice for achieving happiness in the final pages of the *Natural History of the Passions.* Pain and evil come into our lives from the misapprehension of good, and to avoid them we have only to handle our intellects aright. By employing our understanding to examine all things, and by directing our desires only toward those goods which we fully comprehend, we can sidestep the traps of ephemeral and apparent pleasures.

Charleton tried to give his hedonistic ethics the semblance of Christian morality. In a volume from the late years of his life, *The Harmony of Natural and Positive Divine Laws,* he set out to prove that natural or rational morality is equivalent to the moral code revealed in the Bible:

> By nature all wise men understand the order, method, and economy instituted and established by God from the beginning of creation for government and conservation of the

> world. All the laws of nature therefore are the laws of God; and that which is called "natural" and "moral" is also "divine" law, as well because reason, which is the very law of nature, is given by God to every man for a rule of his actions, as because the precepts of living which are thence derived are the very same that are promulgated by the Divine Majesty for laws of the Kingdom of Heaven by our most beloved Lord Jesus Christ and by the holy prophets and apostles; nor is there in truth any one branch of natural or moral law which may not be plainly and fully confirmed by the divine laws delivered in Holy Scripture.[23]

Despite some similarity between the activities recommended by Charleton's hedonism and by Christian morality, basic differences separate the two. Hedonism is egocentric, while Christian love, the element lacking from all of Charleton's discussions of morality, is directed outside the individual toward God and humanity. Hedonism is naturalistic, an ethic for the unaided powers of the individual, while traditionally Christian morality has assumed the necessity of divine grace supporting human intentions. In reducing the attainment of virtue to the intellectual judgments of good and evil, Charleton overlooked the problem of redemption with which Christian theology and Christian practice had wrestled through the centuries. As a Christian he accepted the supernatural role of Jesus Christ, but his works on natural religion reduced religious practice toward the level of reverence for an omnipotent being and conformity to a rational moral code.

Dr. John Wilkins, taking up the study and demonstration of natural religion during the latter years of his life, published his conclusions in a number of sermons and in a posthumous volume, *Of the Principles and Duties of Natural Religion* (1675). His purpose was manifest, to overthrow the doctrines of Hobbes in particular and to refute atheism in general. Since he differed from Charleton in publishing a number of religious books the scope of which extended beyond natural religion, it is possible to judge *Natural Religion* in the context of his whole religious thought. A comparison of the book with his earlier *Discourse Concerning the*

23. Walter Charleton, *The Harmony of Natural and Positive Divine Laws* (London, 1682), pp. 8–9.

Gift of Prayer demonstrates how little of the spirituality of Christianity penetrated the studies of natural principles of religion.

Wilkins opened the *Principles and Duties of Natural Religion* with the postulates on which he would build:

> The nature of man (besides what is common to him with plants and brutes) consists in that faculty of reason whereby he is made capable of religion, of apprehending a Deity, and of expecting a future state of rewards and punishments. Which are capacities . . . which no other creature in this visible world except man partakes of. . . . The happiness of man consists in the perfecting of this faculty; that is, in such a state or condition as is most agreeable to reason, and as may entitle him to the divine favor and afford him the best assurance of a blessed estate after this life.[24]

In Wilkins' opinion men can know the three basic principles of religion by reason alone without the help of revelation. Belief in the existence of God, comprehension of His attributes and perfections, and suitable demeanor toward Him—these three basic principles make up natural religion. Wilkins' treatment took them up in order, considering them not merely as intellectual truths but as basic principles of human happiness.

To prove the existence of God he advanced four arguments: the universal consent of men in all times and places; the finite age of the world; the creative skill displayed in natural things; and the works of providence in the government of the world. His demonstration of the Deity's attributes followed the reasoning that the Supreme Being must contain all perfections. Religious duties were found by consideration of the Almighty's being and excellency which should move us to revere, love, trust, and obey Him; recognition of our own self-interest should reveal the same duties. Wilkins concluded that the nature of man is such that his happiness is found in the practice of religion culminating in the reward of eternal bliss. Part two of the volume upheld the wisdom of practicing the duties of natural religion. As the ultimate happiness and highest perfection of man is attained through the development of his religious faculties, so his happiness even in this

24. Pages 18–19.

world is promoted by the fulfillment of religious duties. Morality is conducive to health; regularity and moderation strengthen the body, while debauchery wrecks it. Whereas neglect of duty by the profligate undermines liberty, safety, and quiet, religion protects and advances these blessings. Moral living produces worldly riches—defined as sufficient means for our wants, and satisfaction and peace of mind with the sufficiency. By making us diligent and reducing extravagance, religious practice increases our means, and by fostering composure of mind it helps us to be satisfied with little. Religion contributes to human pleasure, allowing the fulfillment of proper desires. It enhances the honor and reputation of those who obey it. Finally Wilkins asserted that inward happiness is also increased by the practice of religion, since it regulates the turbulence of passions and fosters tranquillity and peace.

At the end of the book Wilkins hastened to state that natural religion does not detract from the blessings and benefits of the Christian religion, but rather prepares the way for a doctrine agreeable to natural light. Whatever has been said of natural religion, he concluded, "it cannot be denied but that in this dark and degenerate state into which mankind is sunk there is great want of clearer light to discover our duty to us with greater certainty and to put it beyond all doubt and dispute what is the good and acceptable will of God, and of a more powerful encouragement to the practice of our duty by the promise of a supernatural assistance and by the assurance of a great and eternal reward. All these defects are fully supplied by that clear and perfect revelation which God has made to the world by our blessed Saviour." [25] Nevertheless *Natural Religion* took a long step toward reducing Christian morality to a low plane of prosaic self-interest. It argued that the law of self-preservation is the first law of nature; to aim for our own well-being is not only permitted as lawful and commanded as duty, it is necessary to our very nature. In trying to prove that worldly happiness as well as eternal bliss is the reward of virtue, Wilkins was taking the sting out of religious duty and reconciling it to the normal standards of social behavior.

In a sermon from the period when he was meditating on natural religion Wilkins criticized the saint who wished to be cast into

25. Ibid., pp. 394–5.

hell so that he might stop up the mouth of it. Likewise he censured Anselm, who preferred hell to sin and would rather have been in hell without sin than in heaven with it. "Now, I say," Wilkins commented, "though it is possible that such kind of speeches may proceed from those that are good, in some special zeal and heat of their affections, yet they are not from a serious considerate judgment, but being compared with the rule will be found to be unwary boasting expressions pretending to a higher degree of sanctity than men are capable of or than is required from them." [26] No one would deny the reasonableness of Wilkins' words, as no one would care to stop up the mouth of hell, but some element of Christian struggle and striving has been left out. The general tendency of *Natural Religion* was to make religious practice not only rational but reasonable, to replace the pursuit of excellence with a standard of mediocrity, and to substitute for spiritual growth a code of social conduct.

When *Natural Religion* is placed beside Wilkins' *Discourse Concerning the Gift of Prayer,* the difference in religious expression is striking. Where *Natural Religion* was based on the concept of worldly self-interest, the discourse on prayer built from the statement that the "true happiness of every Christian does properly consist in his spiritual communion with God." [27] In prayer the Christian draws on the forgiveness and strength of God to war against his sins and to grow in spiritual perfection. The purposes of *Natural Religion* and the *Gift of Prayer* were different, of course, and the difference of purpose helps to explain the difference of concepts. *Natural Religion* was a form of advertising for religion which attempted to popularize it by proving how reasonable it was. Wilkins himself considered it only as an introduction to the deeper mysteries of Christianity. Nevertheless, the fact that Wilkins took up the exposition of natural religion at all is significant, for it indicates a readjustment of emphases in religion. The religious history of the virtuosi is symbolized in the order in which his books appeared. Instead of the *Gift of Prayer* following *Natural Religion* and adding the teachings of revelation to the

26. John Wilkins, *Sermons Preached upon Several Occasions* (2d ed. London, 1701), p. 33.

27. John Wilkins, *A Discourse Concerning the Gift of Prayer* (London, 1690; first published in 1651), p. 1.

discoveries of reason, it preceded *Natural Religion;* Wilkins progressed from revealed Christianity to natural religion. Like Wilkins the virtuosi gave increasing emphasis to natural religion without denying or abandoning Christianity, until the principles of natural religion became their chief religious expression.

Although Robert Boyle did not publish a systematic discussion of natural religion, much of his scientific work was devoted to it. In book after book he demonstrated the basic principle of religion, bringing his vast knowledge of natural philosophy to the proof of God's existence. Contemplation of the Almighty's excellency as displayed in His works was easily the dominant theme of Boyle's voluminous writings, but he made it perfectly clear that he did not consider the demonstrations of natural religion as the summit of religious truth. Natural theology may achieve a considerable knowledge of God, but there are limits which it cannot transcend. Since God knows Himself infinitely better than human beings can, we must go to His word to augment the lessons of natural theology. Boyle likened the Bible's place in religion to the telescope's use in astronomy. Although we can survey the heavens with the naked eye, a telescope makes everything more clear and reveals much that cannot be seen without it. He thought that natural religion, limited though it is, is nevertheless an excellent introduction to Christianity. Man can recognize that he has been endowed with reason and fitted to discover that there is a Creator both wise and powerful, that he has duties to God under natural religion, and that his soul is immortal. When he considers these things, he will wish that God had given a supernatural revelation of the worship and obedience demanded, a revelation of things unknowable by reason alone. Thus natural religion will lead a man to the threshold of Christianity and prepare him to consider the evidence of the Scriptural revelation.[28] The Bible received more than formal acknowledgment from Boyle. In order to study it more thoroughly he learned the Greek, Hebrew, and Chaldean tongues and read the Scriptures in their original languages. Every reference to the Bible indicated clearly the reverence in which he held it. Boyle never considered natural religion a substitute for Christianity.

28. Boyle, *Works, 5,* 521–2.

Yet Boyle was absorbed in natural religion, deeply absorbed. Unconsciously he allowed it to govern his whole religious outlook. In the study of nature he had seen the power and wisdom of the Creator, and the image of an omnipotent Being possessed his mind. While he acknowledged God the Redeemer, it was God the Creator and Governor of the universe Whom he worshiped. Even such works as *Some Considerations Touching the Style of the Scriptures* (1661) and *The Excellency of Theology* (1674) were dominated by the figure of the Almighty Lord of creation. Since God has implanted notions and principles in the mind of man fit to make him sensible that he ought to adore the Divine Author of his being, Boyle declared in *The Christian Virtuoso,* "natural reason dictates to him that he ought to express the sentiments he has for this Divine Being by veneration of His excellencies, by gratitude for His benefits, by humiliation in view of His greatness and majesty, by an awe of His justice, by reliance on His power and goodness, when he duly endeavors to serve and please Him, and in short by those several acts of natural religion that reason shows to be suitable and therefore due to those several divine attributes of His which it has led us to knowledge of." [29] In the *High Veneration Man's Intellect Owes to God* (1685) he stated that study of God's attributes reveals a glorious majesty "that requires the most lowly and prostrate venerations of all the great Creator's intelligent works." [30] Looking upon the Bible less as a record of God's dispensation to man than as a further and higher revelation of His omnipotence, Boyle centered his religion on awe-struck and even fearful worship of the Creator's power.

The scarcity of references to Christ in the many pages of Boyle's disquisitions on religion is striking. While Boyle manifestly accepted the divinity and redeeming sacrifice of Christ on an intellectual plane, the concept of redemption did not exercise a great influence on his mind. To the attributes of the Creator Boyle devoted endless space; they were the religious questions that exercised his mind. He mentioned Christ, but he discussed God the Creator. The idea of redemption did not play an important role in his thought on moral living; he considered the at-

29. Ibid., p. 520.
30. Ibid., p. 146.

tainment of virtue as an intellectual choice dependent upon a person's comprehension of God. In *The Christian Virtuoso* he concluded from his observations of men that most human perversity results from a hesitant conviction of the existence and attributes of God. It is hard to conceive, he said, "that deliberate, avowed, and habitual impieties and vices can consist with a deliberate and firm belief that there is such a being as even natural religion represents God to be; that is, not an undiscerning, weak, or unconcerned, and in a word a mere nominal Deity, but such a powerful, wise, just, and active author, upholder, and sovereign Governor of the world as philosophy itself has been able to represent God to be." A man's virtue will be proportionate to the firmness of his assent to the fundamental article of religion—that there is a Divine Author and Ruler of the world.[31] In a work from his youth, a didactic "letter" on *Seraphic Love,* Boyle extolled the superiority of love of God over carnal love. To read the piece is to get the impression that choosing between seraphic and carnal love is like judging the relative merits of two books, an intellectual problem merely of deciding which is superior. The same tendency to intellectualize religious practice pervades all of his devotional writings. Apparently Boyle did not keenly feel the need for divine redemption in his own heart. The even tenor of his life was seemingly unbroken by violent temptation, and he pursued a steady course of virtue unmarred by even the faintest breath of scandal. The problems of morality were not pressing concerns in his personal life. He talked about sin like an American discussing cricket; he had heard about it but had never seen it close at hand. The Christian doctrine of redemption rang no response in his soul.

Boyle's lack of concern with redemption did not mean indifference to morality. He always held that Christianity is a rule of life which should regulate and improve the actions of those who profess it. He once quoted the words of Pythagoras that two things must ennoble a man—to know truth and to do good.[32] The dictum characterized his own life as well as his concept of Christian

31. Ibid., 6, 758.
32. Ibid., 2, 5.

practice. Never involved in sectarian disputes and always professing the broadest tolerance, Boyle believed that Christians should live together in perfect charity and devote their energies toward observing God's law. His idea of the moral law had the distinctive flavor of his concept of God. As Boyle thought of it, the moral law is not a path to spiritual growth but a rule governing action like the mechanical laws of physical nature. Obedience is a duty required by the Almighty Creator, and its reasonable fulfillment will end in reward much as a clock that runs properly will trip the alarm. Morality in Boyle's conception is not infused with saving grace; it is the morality of natural religion, obedience to the Lord Who gave laws to all of His creation. Thus his idea of morality brought him back to the conception of Almighty God toward which his religion pointed from every angle. Boyle worshiped a Being shorn of the redemptive mercy distinctive of Christian theology. In the hands of men who were less devout his religious expressions could readily have been turned into deism.

John Ray's religious writings focus on the same image of omnipotent creativity. The study of nature revealed to Ray the wisdom and power of God, wisdom and power excelling His creatures' faculties in infinite degree; and before the mighty figure of God, Ray bowed in meek submission. "Because of His greatness and excellency," he wrote in *Three Physico-theological Discourses,* "He is to be worshiped and adored with the most submissive humility and veneration, with a transcendent and incommunicable worship and devotion."[33] Such a worship of God was not an innovation in Christianity, of course, and Ray left ample testimony that he accepted the doctrines of Christianity that rise above natural religion. Nevertheless, the distinctively Christian doctrines of religion did not move him to write; his religious works were concerned with aspects of natural religion. In the *Wisdom of God Manifested in the Works of the Creation* he attacked atheism with the evidence of creative wisdom found in nature. *Three Physico-theological Discourses* considered the means by which the Author of creation provokes cosmic changes. *A Persuasive to a Holy Life* carried the battle against disbelief into the

33. Page 417.

practical field of daily living. While he believed in Christianity, Ray devoted his religious writing to natural religion and helped to increase its growing importance.

In 1700 Ray published *A Persuasive to a Holy Life.* While the *Wisdom of God* attempted to refute theoretical atheism, the *Persuasive* devoted itself to practical atheism, immoral living which denies God implicitly. Since the virtuosi considered immorality to be the inseparable companion of disbelief, Ray was completing the argument begun in the *Wisdom of God* when he wrote the *Persuasive.* The little book is illustrative of the moral thought of natural religion. Few men obtain happiness, Ray stated in opening, because they pursue chimeras. They look for happiness in bodily pleasures, honors, and riches, whereas it is found only in holy living. Physical and temporal pleasures are deceitful and fleeting. They are limited as well and unable to fill the vast desires and capacity of the soul. The supreme object of man's desire is God; and happiness, which is enjoyment of the object of desire, is found in the knowledge and love of God. That is to say, happiness consists in the holy life, for we show our love of God in our obedience to Him. In the brief introduction the *Persuasive* dwelt on a high ethical plane, a plane too high to persuade those practical skeptics at whom it was aimed. To lure them into the fold Ray descended into the street in the main body of the work to prove that holy living produces the goods of this world as well. Although he had earlier rejected worldly pleasures as deceitful traps, he now admitted that they make up a substantial part of happiness. If he could prove that holiness and obedience to God are the most successful means to procure these enjoyments, he would recommend the holy life to all sorts and conditions of men.

Ray's argument was borrowed, with full acknowledgment, from the last section of Wilkins' *Natural Religion.* The happiness of the external man is built from a number of things. Health, safety, liberty, quiet, riches (defined, following Wilkins, as sufficient means for one's station and a mind contented with the sufficiency), honor, reputation, and friendship—holy living supplies all the ingredients of worldly happiness. Bodily pleasures are not forbidden; only excesses are banned, but excesses have ceased to be pleasures anyway. Internal happiness also springs from the holy life, for virtue brings peace and tranquillity, while the vicious man is never at ease with the torments of conscience and the dread of eternal damnation. The *Persuasive to a Holy Life* completed the domestication of Christian virtue. It presented a morality with which one could live comfortably without inconvenience. Virtue was equated with the solid bourgeois values of respectability, moderation, and caution in all things.

"A holy life and conversation here," Ray said in conclusion, "secures to us an interest in a future state of eternal bliss and happiness, glory and immortality in the world to come, and thereby delivers us from the fear of death . . ." [34] The choice of words indicates the level to which the *Persuasive* reduced morality. Holy living "secures us an interest" in heaven. Religious practice was likened to a commercial transaction; eternal bliss can be purchased by reasonable monthly installments of holy living. When he spoke of Christ in this book, Ray stated that His sacrifice had purchased heaven for men if they are obedient on earth. Christ's death had established a contractual situation of which a man can take advantage through obedience. A man is a fool to let such an opportunity pass. In the final pages of the *Wisdom of God* Ray also included some thoughts on the advantages of religion, which he quoted from Wilkins and which were in common use at the time. A man who denies there is a God and leads a profligate life loses everything if he is wrong. On the other hand a man who governs his life on the assumption that there is a God, suffers only the inconvenience of some needless restraints if he is mistaken. "To which I shall add," Ray continued, beyond the quotation from Wilkins, "that he not only suffers no damage but reaps a considerable benefit from this mistake; for during this life he enjoys a pleasant dream or fancy of a future blessed estate, with the thoughts and expectations whereof he solaces himself and agreeably entertains his time, and is in no danger of being ever awakened out of it and convinced of his error and folly, death making a full end of him." [35] The third physico-theological discourse, on the dissolution of the world,

34. *Persuasive,* p. 116.
35. *Wisdom of God,* p. 464.

concluded with a similar exhortation to holy living. There are sufficient motives to induce anyone to resist temptation, it declared, "such are certain shame and disgrace, and that not long to come, eternal infamy and dishonour, present death, strong fear and dread of approaching death, or sad and intolerable pains or calamities." [36] Morality preached in the language of natural religion became uninspired conformity to an external code recommended by a blunt appeal to self-interest.

Ray and his religious feelings cannot be fairly judged by the *Persuasive to a Holy Life*. All contemporary testimony about him agrees on a sincerity of religious purpose and purity of character unmarred by the complacency that the *Persuasive* would suggest. He won the love and admiration of the worthiest men of his day. A few short prayers that he composed and some passages at the end of the *Wisdom of God* reveal a depth of spiritual insight so superior to the *Persuasive* that they belong to a different genus.[37] The flatness of the *Persuasive* derived from its purpose. It belonged to the class of books on natural religion that sought to extinguish disbelief by proving the rationality of religion. A form of popular propaganda, it was meant to persuade people of the value and validity of religion without making any pretense at profound analysis. As with Wilkins, however, the very fact that Ray wrote the book is indicative of the religious atmosphere in which he lived. The *Persuasive to a Holy Life* was a concrete example of the growing indifference to spiritual problems that accompanied the progress of natural religion.

Among the virtuosi the compulsion to bring natural philosophy and reason to the support of religion was widespread. Joseph Glanvill published in 1671 *Philosophia Pia, or a Discourse of the Religious Temper and Tendencies of the Experimental Philosophy of the Royal Society*. This essay contended that natural philosophy both refutes atheism by proving the existence of God and destroys superstition and enthusiasm by unmasking their groundless imaginings. Even Glanvill's *Saducismus Triumphatus*, which appears to the 20th century to be the antithesis of rational analysis, was considered by its author as a facet of natural religion. Glanvill

36. Ray, *Physico-theological Discourses*, p. 446.
37. *Memorials of John Ray*, ed. Edwin Lankester (London, 1846), pp. 57–62.

believed that repudiation of witchcraft was the product of atheism; and he feared that the mechanical philosophy, which he accepted himself, might be perverted to deny the existence of spirits. If he could employ the tools of careful investigation to prove that witches do exist, he would vindicate the belief in spiritual beings as a whole and thereby strengthen the foundation of Christianity. There had been many impostures about witches, he agreed, but one fully validated story would compensate for a million impostures and suffice to establish the point.

Nehemiah Grew's *Cosmologia Sacra* varied from the norm by directing itself to the refutation of both atheism and deism. After a brief confutation of mechanical atheism Grew devoted most of his book to proving that natural religion is not enough. It must be supplemented by revelation. He intended to demonstrate, he said in the preface, "that there is nothing contained in the Holy Scriptures concerning God or man, the visible or invisible world, but what is agreeable unto right reason." [38] Spinoza more than Hobbes was Grew's *bête noire;* he wrote *Cosmologia Sacra* as a reasoned defense of the Bible's truth and authority.

Thomas Sydenham's short essay on natural religion also differed somewhat from the writings of other virtuosi. Apparently his purpose in composing *Theologia Rationalis* was solely the urge to justify morality on rational grounds. He did not indicate that he considered it to be a rebuttal to theoretical atheism. "The question is," the opening sentence stated, "how far the light of nature, if closely adverted to, may be extended toward the making good men." [39] In the order of nature he found proof of God's existence and grounds for the first duty of natural religion, which is to revere and adore Him. The second duty, arising from our dependence upon God, is to pray to Him for protection and aid. As members of humanity we are further obliged to serve the common good. Finally there is an obligation to our own nature to maintain the ascendancy of soul over body. Since man by himself

38. *Cosmologia Sacra,* preface.

39. Sydenham, *Works,* 2, 307. Latham has printed a truncated version of *Theologia Rationalis* that ends abruptly in the middle of a sentence and omits roughly the final fifth of the essay. Two full copies are found among the MSS of the British Museum: Sloane MS 3,828, fols. 162–171b; and Additional MS 6,469, fols. 107b–116b.

is unable to vanquish the desires of the flesh, God has aided his struggle by the promise of immortality. Natural religion could not carry Sydenham beyond this point; but reason had led him to the threshold of grace, and natural religion was accepted as only the introduction to Christianity.

Although William Petty did not publish his meditations, he too pondered the problems of natural religion at length. As far as his miscellaneous papers reveal his motives, both the urge to confute skepticism and the desire to satisfy reason seem to have been present. On the one hand he sought a rational basis for morality to convert the libertines; on the other he tried to work out a reasonable statement of fundamentals to end sectarian strife and to remove irrationalities from the practice of religion. The proposed treatise on the *Scale of Creatures* was directed toward the first goal. By proving that man is a pitiful creature in the universal scale, his treatise, Petty thought, would "sufficiently humble man and check the insolent skepticisms which do now pester the world, and [be] a good caution against the slighting of religion and practice of good men." [40] A number of papers attempted to fulfill the second goal. Petty was disgusted with sectarian squabbles. When religion becomes absorbed in the discussion of theological minutiae, it forgets that its primary function is to improve the lives of men. "What is vulgarly meant by worshiping, honoring, and glorifying God?" he asked in a paper headed "Religion." Enriching, obeying, and fighting for those who pretend to be His priests, he replied; praying to God for such benefits as man cannot give us; and setting forth God's power, wisdom, and mercy in a general way by metaphorical expressions and allusions when we either ignore or deny them in particulars. From these mistaken notions many disadvantages have flowed—the unnecessary expense of a luxurious church establishment, useless study of inexplicable matters, wars and destruction, superstition, hypocrisy, and inattention to useful arts. God would be more honored by the study and comprehension of natural phenomena.[41] Another paper, entitled "Religio Catholice Catholica" listed nine points

40. *Petty-Southwell Correspondence*, pp. 9–10.

41. William Petty, *The Petty Papers. Some Unpublished Writings of Sir William Petty*, ed. Marquis of Lansdowne (2 vols. London, Constable, 1927), *1*, 117.

of "natural and universal religion." Natural and universal religion consists of studying God's works, worshiping Him in proper assemblies, praising Him in suitable ceremonies, and promoting the welfare of other men. If anyone believes in any further doctrines, let him go beyond the public worship in his private family.[42] In the same vein Petty suggested in another paper that no one should interfere with the religion of others if it fosters moral uprightness, gives comfort and courage, and promotes good citizenship.[43] A declaration of *Twelve Articles of Faith* condensed Christianity into what he evidently considered to be its necessary fundamentals: eternal, omniscient, and omnipotent God created the world and all of its creatures; the soul of man is immortal; God rewards those who seek Him and punishes the wicked; Jesus of Nazareth, His son, came to redeem fallen man; to "do as we would be done unto is fulfilling of the law, the prophets, and the will of Christ." [44] A large number of Petty's papers dealt with religion. They were unified by their concern with a problem which obviously troubled him—the discovery of basic comprehensible principles of devotion that would both satisfy rational men and provide a reasonable statement of common beliefs on which all Christians could unite.

> Religion's natural, and good
> For king or state, if understood;
> If not, 'tis but a mere illusion,
> Begetting bloodshed and confusion.[45]

These were the words of natural religion.

William Petty marks a turning point in the virtuosi's consideration of natural religion. The words of his ditty—"Religion's natural, and good / for king or state, *if understood*"—introduce a spirit that was not present in the works of Wilkins, Boyle, or Ray. Where they had brought reason to the support of religion, using it to demonstrate fundamentals, Petty was beginning to employ reason as a tool of criticism. Once the validity of reason in re-

42. Ibid., pp. 130–1.
43. Ibid.,.p. 118.
44. Ibid., pp. 131–2.
45. Ibid., 2, 251–2.

ligion was admitted, it was difficult to check its sphere. With some of the virtuosi the demand for comprehensibility, already strong in Petty—the demand that religion not only rest on rationally demonstrable foundations but also teach doctrines plainly understood—became all important. When this step was taken, natural religion ceased to support Christianity and began to replace it.

The reports of Edmond Halley's religious views suggest that he had fully abandoned Christianity. It is impossible to determine Halley's opinions finally. Rumors and charges against him have survived, but little solid evidence that can be checked. One fact is certain: in 1691 Oxford University rejected his application for the Savilian chair in astronomy because of some doubt about his orthodoxy, although the exact charge is not known.[46] There is no evidence on which to convict Halley of atheism, and passages in his scientific writings praising the skill of God certainly imply that he believed in the existence of a Creator. Although he was probably a theist, he was evidently not a Christian if two stories can be believed. To his religiously minded friend Robert Nelson, who was arguing against atheism, he supposedly replied that he agreed with his proofs of the existence of God; and then he added, "But by God if you think to bring me in for any more of the family, you are mistaken." [47] Thomas Hearne recorded that Halley told Bishop Stillingfleet that he "believed a God and that is all." [48] The many charges against Halley probably had some foundation; the two stories, which suggest that he was a deist, may be untrue, but they are not unbelievable. In rejecting the Trinity, Halley would only have actualized the tendency of other virtuosi to exalt the Father above the Son. It was a position defensible by

46. In a letter to Abraham Hill, June 22, 1691, Halley said that he was accused of asserting the eternity of the world (*Correspondence and Papers,* p. 88). Perhaps this involved the broader charge of atheism.

47. Bodleian Library, Rawlinson MS J 4, fols. 103, 103b, 105, 105b; cited in Eugene Fairfield MacPike, *Helvetius, Flamsteed, and Halley; Three Contemporary Astronomers and Their Mutual Relations* (London, Taylor and Francis, 1937), pp. 72–3.

48. Hearne, *Remarks* . . . (11 vols. Oxford, 1885–1921), 3, 472–3; cited in Eugene Fairfield MacPike, "Halley, Flamsteed, and Newton," *Notes and Queries, 168* (1935), 436–7.

reason, while the Trinity was an incomprehensible mystery founded on revelation.

John Locke was more typical of the virtuosi. In his works natural religion continues to clothe itself in the traditional terminology of Christianity. On the surface they appear to follow the pattern of employing natural religion as a rational defense of Christianity's foundations. In the *Essay Concerning Human Understanding* (1690) he presented proofs of the existence of God, the immortality of the soul, and the moral obligation of natural law; and he developed a theory to admit the validity of revealed doctrines above reason. *The Reasonableness of Christianity* (1695) brought reason to the support of what he considered to be fundamental Christian doctrines. Locke's purpose, however, differed from Boyle's. He was endeavoring not to defend Christianity from the growth of atheism but to prune it of every doctrine not susceptible to rational demonstration or corroboration. While he admitted in theory that religion might accept doctrines above reason, in practice he did not leave any significant beliefs to fill that category. *The Reasonableness of Christianity,* far from upholding the reasonableness of traditional Christian doctrines, interpreted Christianity in such a manner that it eliminated everything above common sense. The common fallacy that reason is not to be admitted in matters of religion, he asserted, has led to all manner of absurdities and superstitions. "So that in effect religion, which should most distinguish us from beasts and ought most peculiarly to elevate us, as rational creatures, above brutes, is that wherein men often appear most irrational and more senseless than beasts themselves. . . . 'I believe because it is impossible' might in a good man pass for a sally of zeal, but would prove a very ill rule for men to choose their opinions or religion by." [49]

In the preface to the *Second Vindication of the Reasonableness of Christianity* Locke mentioned the reasons that had prompted him to write the treatise he was now defending. One day when he was thinking about the question of justification, which raised

49. John Locke, *An Essay Concerning Human Understanding* (London, 1690), Bk. IV, chap. xviii, Sec. 11.

so much noise and heat among the dissenters, he decided to examine it more fully. The Scriptures made it clear that faith justifies, but the meaning of "faith" was a matter still to be determined. To decide it, Locke searched the New Testament; the *Reasonableness of Christianity* embodied his conclusions. Seen in this light, the work appears as the attempt of reason to resolve the debates of sectarians and to dissolve the obscurities of enthusiasts, an exercise in religious rationalism. He also gave a second reason for composing the treatise—the desire to refute the contention of deists that Christianity contains doctrines beyond comprehension.[50] Both motives looked toward the removal of seeming irrationalities from Christianity.

To understand the doctrine of redemption, Locke began, we must understand what was lost by Adam's fall. By his fall from a state of perfect obedience, which is called justice, Adam forfeited paradise and immortality. Death, as the inheritance of Adam, does not, however, signify eternal torment in hell, nor does it mean that man must necessarily sin. It means simply that man became mortal. After Adam's fall man was to live by the law of reason, and exact obedience would merit the immortality that Adam had lost. Since there was no remission of sins, one false play lost the whole game, and no one was able to win eternal life. God therefore sent Jesus into the world to establish a new set of rules—the law of faith to replace the law of works. Faith is allowed to supply the defect of obedience, and believers are admitted to immortality as if they were wholly righteous. In Locke's reading of the New Testament the law of faith requires that everyone, as a condition of the covenant with God, shall believe what God requires him to believe—that there is one eternal and omnipotent God and that Jesus was the promised Messiah. Further, a believer must repent and live in sincere obedience to Jesus' law to the best of his powers. Locke maintained that the only article of faith stressed in the New Testament is that Jesus was the Messiah. Belief in Jesus, accompanied by sincere repentance, suffices to win salvation. What happens, it was asked, to those people who have never heard of the Messiah? Locke replied that God levies duties in proportion to privileges and oppor-

50. Locke, *Works, 6,* 186–9.

tunities. Since many men have never received the revelation of the Messiah, they are not in a position to accept or reject it. Because God has revealed His goodness and mercy to all men who make use of their reason, the light of reason, which reveals the law, should also reveal the means of conciliating the merciful Father.[51]

In the vindications of the *Reasonableness of Christianity*, which Locke published in reply to attacks, he insisted that he had tried to present not a complete Christian theology but only the essentials necessary for salvation. There is a faith that makes a man a Christian, and that faith is the belief in Jesus as the Messiah, the promised deliverer. There may be other doctrines which are perfectly true, but faith in Christ is enough to win forgiveness and eternal life. Nevertheless, in raising the one doctrine above the others, Locke was in effect denying the others and reducing Christianity to one comprehensible and reasonable belief. His understanding of redemption bore no resemblance to the traditional doctrine of inner regeneration. Significantly he adopted the Thomist definition of faith—belief in a revealed doctrine—and rejected the Protestant notion of filial reliance. With the concept of atonement in its traditional meaning separated from his notion of Christ's mission to the world, Locke had no reason to consider Christ to be a part of the Godhead. Locke was accused of Socianianism both by Bishop Stillingfleet, who objected to the *Essay Concerning Human Understanding*, and by John Edwards, who attacked the *Reasonableness of Christianity*. To both charges Locke replied with evasions which, by avoiding direct affirmation of his belief in the Trinity, left no doubt that he did not accept it.[52] If the function of Christ in salvation is no more than Locke concluded, why was He sent at all? Locke suggested five reasons. Christ came to confirm the existence of the unitary omnip-

51. Ibid., 6, 132–3.

52. Locke, "First Letter to the Bishop of Worcester" (ibid., 3, 96); "Second Reply to the Bishop of Worcester" (ibid., 3, 200–1); "Vindication of the Reasonableness of Christianity" (ibid., 6, 163–7). *Cf.* "Reasonableness of Christianity" (ibid., 6, 22, 32–3); "Adversaria Theologica," cited in Lord Peter King, *The Life of John Locke, with Extracts from His Correspondence, Journals, and Commonplace Books* (new ed. 2 vols. London, 1830), 2, 186–94; Bodleian, Locke MS f 30, fol. 42.

otent God, to establish the moral law more firmly, to reform the outward mode of worship which was filled with excessive pomp, to encourage virtue and piety by revealing their reward more clearly, and to give the promise of assistance to the virtuous life.[53] The fifth reason bore the germ of the traditional notion of salvation; Locke never developed it. As his critics contended, he had departed radically from orthodoxy. Although he still used the name "Christianity," the differences separating his religion from deism were essentially semantic. Natural religion had displaced Christianity almost completely in his thought.

Beside the one article of faith Locke placed the necessity for repentance. To gain salvation a man must repent his past sins and endeavor to live comformably to God's law in the future; belief in Christ will redeem the occasional lapses of those who sincerely strive for virtue. In accordance with the usual emphases of natural religion Locke placed far greater stress on morality than on correct belief. The Christian religion, he wrote, "is not a notional science to furnish speculation to the brain or discourse to the tongue, but a rule of righteousness to influence our lives . . ."[54] Just as he attempted to set theology on a rational plane, so he believed that the moral law could be discovered by reason. Given the idea of God, infinite in power, goodness, and wisdom, and the idea of man as a rational creature, morality should be subject to as accurate demonstration as mathematics. He added that in actual practice men were not wholly successful in uncovering the natural law, but the coming of Christ revealed it completely. The Christian moral law is identical with the natural law.

Locke was evidently dissatisfied with the purely rational ethic. While it might suffice for philosophers, the ordinary man required an ethic with sanctions. In Book II of the *Essay Concerning Human Understanding* he turned away from pure rationalism to adopt hedonism. He argued that God has attached the sensations of pleasure and pain to various experiences as spurs to human

53. Locke, *Works*, 6, 138–51.

54. The creed composed by Locke for the Society of Pacific Christians; cited in H. R. Fox Bourne, *The Life of John Locke* (2 vols. New York, Harper, 1876), 2, 185.

action, and that He has further established future rewards and punishments as additional incentives. Men constantly pursue happiness by seeking to maximize their pleasure. Since the secret of true happiness is to forego ephemeral pleasures in order to win lasting ones, and since eternal life is the greatest pleasure of all, every action should be subordinated to its achievement. Locke's hedonism was a compromise with rationalism rather than a rejection of it. In his opinion reason was to weigh pleasures against each other and direct the will along the path to real happiness. The hedonistic ethic reinforced his abandonment of the traditional Christian concept of salvation. Gone and forgotten were the notions of redemption, regeneration, and spiritual growth. Cool calculation displaced religious fervor as the mainspring of moral activity, and ultimate salvation became a matter of carefully adding pleasures and pains to balance the books in the black. It was religion made reasonable to the counting-house clerk.

If in theory Locke's ethics undermined the Christian concept of salvation, his practical notion of the good life cut asceticism out of morality. An early composition, *Reflections upon the Roman Commonwealth,* praised the Roman religion, which, he noted, was not "soured with needless severities and affected austerities by imposing doctrines of penance, abstinence, and mortification, which serve only to cross the innocent appetites of mankind without making them better or wiser." [55] Like John Ray's *Persuasive to a Holy Life,* Locke's writings tended to domesticate Christian ethics. Three letters of advice sent to Denis Grenville in 1677 and 1678 summarized his idea of moral living. Grenville was troubled by the necessity he felt to do the best thing at every minute, and he was finding the burden intolerable. Locke assured him that we are not under such an exact law. Certain things are forbidden; they can never be done. Others things are required and must be performed whenever occasion arises. In most things we are left the freedom to choose within a range of permissible actions without fear of damnation. If we had always to do the best, we should never finish debating what is the best. In the final letter he asserted that worldly business and devotion are not

55. Cited in ibid., *1,* 150.

contradictory. Since we are born with a body, we must sustain it; the whole day cannot be spent in halleluiahs.[56] The letters are full of excellent advice which is rational and acceptable to modern man. Indeed its significance lies in its acceptability to modern man. Christianity was reconciled to the world, and Locke prescribed a course of religion which a man could endure without discomfort. Spiritual rigors gave way to urbanity, which was tolerant, charitable, sincere, and even noble but was still urbanity for all of that.

In practicing religion, Locke revealed a purity of character and a depth of spiritual insight worthy of the best tradition of the Christian religion. In writing about religion, however, he sacrificed worship of the heart to charity of intellectual concepts. "Religion," he wrote in his Journal, "being that homage and obedience which man pays immediately to God, it supposes that man is capable of knowing that there is a God and what is required by and will be acceptable to Him thereby to avoid His anger and procure His favor." [57] The words mark the culmination, as it were, of the religious development of the virtuosi. Pithy and precise and shorn of obscurities, they squeeze religion neatly into a rational category—sacrificing in the process nothing more than the juice of life.

Not all of the virtuosi participated in the movement toward natural religion. John Wallis, an orthodox Calvinist to his dying day, thought that it was unwise to quit the principles of Christianity and to rely upon reason and natural light. Much of Christianity depends upon revelation; and though it is not repugnant to reason, it is much above reason and not able to be discovered by reason without the aid of Scriptural revelation. "We do but too much gratify men of atheistical and unchristian principles," he asserted, "when (to comply with their cavils) instead of holding fast what is good we let go our hold and dispute our principles, as if we were now planting Christianity among heathens and not edifying a Christian church." It is good to have solid foundations, he continued, but let us not be strengthening them forever. Sometime we must build upon them, and surely they are

56. Printed in ibid., *1*, 388–97.
57. Bodleian, Locke MS f 5; entry for April 3, 1681.

solid after fifteen hundred years. "If out of a needless scrupulosity to satisfy the cavils of those who do not desire to be informed we be always digging at the foundation upon pretense of searching it, we do thereby weaken rather than strengthen it, and (like the foolish builder) having laid the foundation, shall never be able to finish it: ever learning and never able to come to the knowledge of truth." The fundamental doctrines of Christianity were settled long ago. "And if any go about to shake the foundations, we are not therefore, in compliance with their humours, to admit for disputable whatever they please to cavil at, but to hold fast the truth, to hold fast the form of sound words, as well settled and long since agreed upon. . . . While we are taking pains to satisfy such (who resolve not to be satisfied) we may sooner raise new scruples in the minds of well-meaning persons than satisfy those who are willfully ignorant." [58]

Wallis' warning was a good prediction of the ultimate effect of natural religion. Instead of becoming the sure foundation of Christianity, it became its rival. Perhaps it would better be said that natural religion promoted a major revision of Christianity, since the virtuosi as a whole did not reject the name of Christian. The test of reason, once admitted, could not be limited to fundamentals, and John Locke applied its cold query to the whole structure of Christian theology, repudiating those doctrines that could not meet the test. In abandoning the true ground of Christianity and arguing religion as though it were natural philosophy, the virtuosi did more to challenge traditional Christianity than all of the "atheists" with whom they did battle.

The whole religious thought of the virtuosi was colored by their concentration on natural religion. The Christianity which they preached shrank down toward natural religion, until it consisted primarily of veneration for God, obedience to the natural law, and belief in an after life. When John Locke virtually reduced Christianity to identity with natural religion, he was only making explicit a development implicit in the writings of others.

In consequence of the attention that they paid to natural religion the virtuosi ignored for the most part some of the traditional problems of Christianity. Their religion became decidedly ex-

58. Wallis, *Resurrection Asserted,* pp. 26–30.

ternalized, focusing more on the manifestations of divine power in the physical world than on the internal realm of the spirit. Although it was far more sophisticated than the miracle-studded hagiographies of the Middle Ages, it rested on the same response of worship and awe for the omnipotent Being Who could push brute matter around. Meanwhile the personal problem of spiritual growth received scant notice. The Christian doctrine of salvation declined in their writings on natural religion to the analogy of an athletic contest. Eternal life was the trophy awarded to those who performed well, and the whole wide question of how a man can be righteous was left unanswered, was scarcely recognized. While no one can reasonably doubt the religious sincerity of the virtuosi and their personal acceptance of Christianity, the natural religion that they developed did not even acknowledge the fundamental problem with which Christianity had dealt in the past, the problem inherent in the anguished cry of the apostle, "O wretched man that I am! who shall deliver me from the body of this death?" Natural religion was a response of the intellect to external facts. It was not immediately related to the concrete spiritual experience of the individual in a manner which made religion an integral part of his life as well as an intellectual creed. In reducing the significance of Christ, in stressing the transcendency of God, and in preaching uninspired obedience to an objective law, the natural religion of the virtuosi prepared the ground for deism.

On the other hand the virtuosi's religion was more than socially conditioned assent to traditional beliefs. There was another aspect of their religion which mitigated its intellectualism to a considerable degree. An intense feeling, shared by many of the virtuosi, of personal contingency and dependence upon divine sustaining power lay at the core of their religious consciousness. Death and suffering were not strangers to the virtuosi. Sir Kenelm Digby lost his beautiful young wife to a disease which would not claim her now, and he himself was tortured by the "stone" for over twenty years. John Wilkins suffered and died from the stone, while Locke was the victim of consumption which left him a semi-invalid during his final ten years of life. John Ray's legs rotted away with an unidentified dermatitis, and Boyle was a conscientious valetudinarian who never allowed himself to enjoy a day of health. As early as the age of twenty Boyle suffered from a renal calculus—"gravel of the kidneys" in the picturesque phrase of the 17th century. Throughout his life he collected remedies which he published in 1688 in a volume called *Medicinal Experiments.* In the preface to the book he listed with loving concern the succession of ailments which had tormented his frame—an index to life in the 17th century, and an indication of the reason why eschatology could not long be neglected in that age. As the fourteenth child of a mother who died of consumption at the age of forty-two he had never been robust, and he probably injured his health by excessive study. His specific misfortunes began in Ireland, where a severe fall from a horse bruised him so that he never ceased to feel the effects. When he was forced to travel before he was fully recovered, he met with bad weather, poor accommodations, and a drunken guide who caused him to wander all night in the mountains. All combined to put him into a fever and dropsy which forced him to go to London for a cure. Arrived in London, he was just in time to catch "an ill-conditioned fever" raging there; as a farewell when he recovered, it threw him into a violent "quotidian" or "double tertian" ague which dimmed his eyesight. There followed a "scorbutic cholic," which deprived him of the use of his hands and feet for many months; and protracted sitting brought on the stone.[59] It was a remarkable tale of woe even for the 17th century. In one of his *Occasional Meditations* Boyle pondered the meaning of sickness. When we are long accustomed to health, he concluded, we take it for granted without realizing what a blessing it is; "we are not sensible enough of our continual need and dependence on the divine goodness if we long and uninterruptedly enjoy it; and by that unthankful heedlessness we do, as it were, necessitate providence to deprive us of its wonted supports to make us sensible that we did enjoy and that we always need them."[60]

Samuel Hartlib fell victim to a sequence of diseases as shattering as Boyle's. The stone, internal piles, an ulcer, and finally an affliction called the "dead palsy" all combined to torment his final

59. Boyle, *Works,* 5, 315.
60. Ibid., 2, 366.

years. With Boyle he found in sickness a pathway to God. "The stone is like a bull enraged that will not fall with one blow," he wrote in pain to his friend Worthington. "But strong is the Lord God Almighty even in this case to save His poor servants that trust in Him alone. His name be blessed for ever." [61] In a similar way the autobiographical sketches and letters of John Flamsteed display a vivid feeling of dependence upon God for the sustenance of life. Flamsteed, like so many others, was victim of the stone. Scarcely a letter that he wrote omitted a phrase such as "if God spare me life and health," or "if God spare me health and time." Imminent death floated constantly before him; he could not doubt the dependence of his life upon God's good grace. Perhaps someone should study the impact of modern medicine upon religion. Death and suffering are not the constant companions of daily life in the 20th century that they were in the 17th. It was impossible for the virtuosi to think that they were self-subsistent when every day death and disease reminded them how slender was the thread that supported life. The personal religion of the virtuosi was founded more on the intuition of individual contingency and reliance upon divine support than on the need for redemption from sin.

Although the doctrines of natural religion omitted the spiritual teachings of Christianity, some of the men who developed natural religion could stand comparison with the purest saint of the Christian church. John Wilkins, Robert Boyle, John Ray, John Locke—they were men of the most refined tempers and ennobled characters, whose lives were infused with spirituality and charity that were profoundly Christian. The type of religious argument in which they engaged dictated the form of their religious writings more than did their own spiritual insight. John Wallis urged them not to acknowledge annoying objections, yet to ignore the objections was precisely what they could not do. They had to examine them to satisfy their own peace of mind. They lived in an age that was questioning accepted standards, and they themselves were leading the greatest movement of questioning. They believed in reason, which discovered so much in natural philosophy. Their refined natures were revolted by the obscurity of

61. *Worthington's Diary,* 2, Pt. I, 68.

enthusiasm. Someone had to apply the test of reason—to turn back the objections, which were too close to their own beliefs to be ignored, and to free religion of parasitical growths, which threatened to choke the whole plant. Someone had to expound natural religion. Not every virtuoso felt the compulsion of reason so strongly that he needed to demonstrate the soundness of Christianity's foundations; those who did feel the compulsion could not deny their own natures and forsake the investigation. More than answering hypothetical atheists, they were trying to satisfy their own doubts; they were trying to regain certainty after a new system of thought had called their most cherished values into question. Although their efforts to defend Christianity concluded by undercutting it, their work was essential to religion; for they examined questions that could not be evaded.

CHAPTER 6

Survivals of Supernatural Christianity

"The will of God," says St. Paul, "is our sanctification." What is that? what, but that the decays of our frame and the defacements of God's image within us should be repaired . . . that, in short, we should become friends of God, fit to converse with angels and capable of paradise.

Barrow, *Of Submission to the Divine Will*

Let us consider that charity is a right noble and worthy thing, greatly perfective of our nature, much dignifying and beautifying our soul. It renders a man truly great, enlarging his mind unto a vast circumference and to a capacity near infinite, so that it, by a general care, reaches all things, by an universal affection embraces and grasps the world.

Barrow, *Motives and Arguments to Charity*

WHILE the general movement of religious thought among the virtuosi was to define and solve the problems of natural religion, three virtuosi were prominent in their refusal to become absorbed in the rational demonstration of religious fundamentals. Sir Thomas Browne, Isaac Barrow, and John Mapletoft did not quarrel with natural religion or object to it, but they considered its conclusions to be unquestioned truths which did not require lengthy consideration. Their eyes focused on other problems, which proofs of the existence of God did not touch. To the semi-mystical vision of nature held by Sir Thomas Browne, Robert Boyle's long investigation of the mechanical skill of God in the creation must have seemed wholly beside the point. Both Barrow and Mapletoft were willing to accept the support of natural religion but not to stop short with its conclusions. The virtuosi who discoursed on natural religion maintained that it was only the foundation of higher Christian truths. Browne, Barrow, and Mapltoft insisted on treating it as such. Where others were content to remain outside admiring the temple of God, they went inside to worship.

Although Sir Thomas Browne (1605–82) stood only on the fringe of the scientific movement, his interest in natural philosophy, centering on the biological sciences, was enough to make it an important, if not the major, influence in his life and thought. His conception of nature diverged radically from the mechanical hypothesis of the other virtuosi, however, and meshed with a religious viewpoint which also contrasted with theirs. "The wisdom of God," Browne wrote in *Religio Medici* (1643), "receives small honor from those vulgar heads that rudely stare about and with gross rusticity admire His works; those highly magnify Him whose judicious inquiry into His acts and deliberate research into His creatures return the duty of a devout and learned admiration." [1] The words might well have been those of Boyle, but appearing in *Religio Medici* they carried a meaning different from similar passages in Boyle's works. The visible world, in the opinion of Sir Thomas Browne, is but a picture of the invisible, with objects assuming equivocal shapes which represent some higher reality in the invisible fabric. Nature is a book written in shorthand, a message in hieroglyphs which serve to intelligent minds as symbols to interpret the truths of divinity. In the philosopher's stone Browne discovered an obscure figure of the perfection and exaltation of the soul when freed from its prison of flesh. The "transmigration" of silkworms, he said, turned his philosophy into divinity. "There is in these works of nature, which seem to puzzle reason, something divine and has more in it than the eye of a common spectator does discover." [2] In preference to Epicurus, Browne took his conception of nature from Plato; and instead of seeing reality in the mechanical conjunction of material particles, he believed that the physical world dimly shadows forth a spiritual reality behind it. In the *Garden of Cyrus* (1658) his mystical approach to nature reached its fullest expression, as Browne discovered representations of ultimate truth in the number five and the symbolic figure of the quincunx. In the words of Coleridge he found quincunxes in the mind of man, quincunxes in tones, in optic nerves, in roots of trees, in leaves, in everything. "All things

1. Sir Thomas Browne, *The Works of Sir Thomas Browne*, ed. Geoffrey Keynes (6 vols. London, Faber and Faber, 1928–31), *1*, 18.

2. Ibid., *1*, 50.

began in order," Browne mused, "so shall they end, and so shall they begin again, according to the ordainer of order and mystical mathematics of the City of Heaven." [3] Here was a conception of nature ideally suited for the religious dreamer.

Browne's method of studying nature, if it may be dignified with the title of "method," surrendered to his religious bent. To probe the spiritual mysteries of nature and to interpret the symbols of eternal truth were the tasks that he set for his philosophical inquiries. On the surface of nature lie the empirical data from which alone truth can be built, but the intellect must seize upon the facts and by an act of the imagination penetrate beneath their outward appearance to their inward nature. Truth, which is said not to seek corners, he wrote in his commonplace book, "lies in the center of things, the area and exterous part being only overspread with legionary varieties of error or stuffed with the meteors and imperfect mixtures of truth." [4] The *Garden of Cyrus* concluded with a statement confirming this approach to truth.

> Flat and flexible truths are beat out by every hammer, but Vulcan and his whole forge sweat to work out Achilles his armour. A large field is yet left unto sharper discerners to enlarge upon this order [of quincunxial symbols], to search out the quaternios and figured draughts of this nature, and moderating the study of names and mere nomenclature of plants, to erect generalities, disclose unobserved proprieties not only in the vegetable shop but the whole volume of nature, affording delightful truths confirmable by sense and ocular observation, which seems to me the surest path to trace the labyrinth of truth.[5]

The "delightful truths" for which Browne searched were not the natural miracles that pleased the other virtuosi so much. While the latter described the mechanical operations of nature, Browne was drawn into the recesses of esoteric profundities, there to speculate on mysteries and spin imaginative delights. In the causes and nature of eclipses there is excellent speculation, he

3. Ibid., *4,* 125.
4. Ibid., *5,* 245.
5. Ibid., *4,* 124.

suggested in *Religio Medici,* but to penetrate more deeply and to contemplate why God's providence has so disposed the motions of heavenly bodies in their vast circles to conjoin and obscure each other is "a sweeter piece of reason and a diviner point of philosophy." [6] Nature remained an unprobed depth of spiritual truth tempting Browne's imagination, an infinite field in which his mind could wander, forever pursuing fresh speculation on the mysteries of divine wisdom.

The same sense of impenetrable mystery underlay Sir Thomas Browne's religion. Indeed his speculations about nature cannot be separated from his religion, for in the formal worship and doctrine of the Church he was probing into the same boundless chasm which he approached through philosophy. Faith came easily to Browne; by his own declaration he found in Christianity too little for an active faith. He agreed that reason poses serious doubts about religious beliefs; but having convinced himself that reason cannot explain even the phenomena of nature, he taught his reason to submit to faith and conquered his doubts, not in a warlike posture, but on his knees. To believe only possibilities is philosophy, not faith, and a nobler act is to believe what reason holds to be impossible. In *Religio Medici* Browne listed some of the doubts and puzzles with which reason beset Christian revelation. Where was the soul of Lazarus while he was dead? Could not his heir have claimed his goods after Lazarus returned to life? And from the chasm the answer returned: "These are niceties that become not those who peruse so serious a mystery." [7]

The greatest mystery of all, and the one that continually exercised his mind, was the mystery of life and death. The subject held a fascination for him; again and again he paused to consider the mutability of the world and the ultimate release by death. The posthumous *Letter to a Friend* (1690) was an opportunity to play on the theme, and the dignified and solemn *Urn Burial* (1658) presented the contrast between worldly mutability and oblivion and Christian immortality. Without ranting or beating his breast, Browne looked on the world as an unbroken display of vanity and foolishness dissolved in the end by death. Only the

6. Ibid., *1,* 20.
7. Ibid., p. 30.

Christian promise of immortality brought reason into this mystery of existence and made the torments of life endurable. "Certainly there is no happiness within this circle of flesh, nor is it in the optics of these eyes to behold felicity," *Religio Medici* intoned. "The first day of our jubilee is death; the devil has therefore failed of his desires: we are happier with death than we should have been without it." [8]

The religion of Sir Thomas Browne was a spiritual response to the spiritual mystery that he perceived around him. Keenly aware of his own contingency and of his inclination to sin, conscious of a higher reality in which the contradictions and enigmas of the material world resolve themselves into truth, he found release in spiritual communion with the source of being and truth. "I am sure there is a common spirit that plays within us, yet makes no part of us," he said,

> and that is the Spirit of God, the fire and scintillation of that noble and mighty essence which is the life and radical heat of spirits and those essences that know not the virtue of the sun, a fire quite contrary to the fire of hell. This is that gentle heat that brooded on the waters, and in six days hatched the world; this is that irradiation that dispels the mists of hell, the clouds of horror, fear, sorrow, despair, and preserves the region of the mind in serenity. Whosoever feels not the warm gale and gentle ventilation of this Spirit, though I feel his pulse, I dare not say he lives; for truly, without this, to me there is no heat under the tropic, nor any light, though I dwelt in the body of the sun.[9]

Mystic or semimystic, worshiping with the wordless devotion of the spirit, Browne did not feel the need to define and demonstrate fundamentals of natural religion. His religion was an act of his whole being, not a set of intellectual propositions; its validity rested on a logic different from the reasoning of natural philosophy, attested by an inner light. Browne hated theological disputes and sectarian quarrels over articles of belief. To him the substance of religion was the life of the spirit; the ceremonies and doctrines

8. Ibid., p. 55.
9. Ibid., p. 40.

of the Church were outward symbols of the indefinable objects of faith just as the phenomena of nature symbolized a hidden reality. Although his flights of inspired imagination rose from the platform of natural religion, he considered its series of propositions so indubitably true as not to be in need of proof. All of the religious problems with which his mind was occupied dwelt on another plane where the objections of mechanical materialism could not impinge on them.

There was an element of skepticism in Browne which challenged traditional beliefs. "In philosophy, where truth seems double-faced, there is no man more paradoxical than myself," he declared; and again, "I perceive the wisest heads prove at last almost all skeptics and stand like Janus in the field of knowledge." [10] When his skepticism is examined, however, it proves to have been a milder form of doubt than the great repudiation of traditional philosophy by the leaders of the scientific movement. In *Vulgar Errors* (1646) Browne exposed a host of superstitions and mistaken beliefs, all of a rather singular and esoteric character. While the leaders of natural science were denying Aristotelian metaphysics and Ptolemaic astronomy, Browne condemned the belief in griffins and cockatrices. His skepticism in philosophy was not of the type that challenged whole world-pictures. Most of the "vulgar errors" with which he dealt were connected in one way or another with Christian tradition, around which they formed a vast body of mythology without Scriptural foundation. *Vulgar Errors* was in part an expression of Browne's religion, an effort to cleanse the spring of truth of the rank weeds that choked and polluted it. His "skepticism" stopped short at this point. He was not the scientific questioner who accepted the Church's articles of belief on faith while he probed and doubted the philosophical principles that supported Christianity. His platonic conception of nature fitted harmoniously into his spiritual view of Christianity. Far from challenging the bases of religion his "paradoxical" spirit was only trimming away superstitions which obscured pure spirituality.

Isaac Barrow (1630–77) was more intimately connected with the scientific movement than Sir Thomas Browne. Although he

10. Ibid., pp. 10, 86.

was not an experimenter, his genius in mathematics helped to prepare the way for Newton's discovery of the calculus. Barrow resembled Browne in his concern with spiritual problems. The urgency of his religious experience finally swamped his interest in mathematics and led him to abandon the study in order to devote himself wholly to the Church. A Latin verse sent with his published mathematical lectures to John Tillotson in 1669, just before he resigned the Lucasian chair of mathematics at Cambridge, reflected his growing impatience to be at work that he considered more important. "While you, my dear friend," he wrote to the practicing clergyman, "deliver the mysteries of sacred truth to the people in powerful eloquence, and shut the mouths of petulant sophists, and carry on successful controversy for the law of God; I, as you perceive, am so unhappy as to be fixed to the books which you see here, and so waste my time and my power." [11] Barrow's renunciation of mathematics as a profession in order to serve the Church spoke of religious yearnings that would not rest content with the devotion that other virtuosi drew from the study of nature. While he could use the materials of natural religion and construct demonstrations of fundamental religious principles, he refused to be satisfied with natural religion. As his letter to Tillotson said, he longed to employ his powers in delivering the mysteries of sacred truth to the people.

Barrow was not opposed to the use of reason in religion; he did not demand blind assent to Christianity. On the contrary, he maintained that the peculiar excellence of Christianity lay in its appeal to reason. Since man is rational, true religion must conform to his rationality, and more than any other religion Christianity invites and requires a thorough examination of its fundamentals.[12] Not only is the basic principle of Christianity, the existence of God, knowable by reason alone, but God is the most intelligible being that exists. What little we know of created beings is only the reflection of His intelligibility; He is like the sun without which we can see nothing.[13] Above the rationally

11. Cited in William Whewell, "Barrow and His Academical Times," *The Theological Works of Isaac Barrow, D.D.*, ed. Alexander Napier (9 vols. Cambridge, 1859), *9*, xxxviii–xxxix.

12. Napier, *5*, 394–410.

13. Ibid., *4*, 478–87.

demonstrable fundamentals lies a body of revealed truths to be accepted on faith once the foundations have been established. This superstructure of Christianity was the focus of Barrow's religious life. He was not much concerned with the demonstration of fundamentals; their truth imposed itself so powerfully upon his mind that he was unable to see how a reasonable man could doubt them. Without objecting to attempts to demonstrate the rationality of religion, Barrow concentrated his own attention upon other religious questions with the result that his sermons and writings bear no resemblance to natural religion.

Where natural religion led to the de-emphasis of some traditional doctrines of Christianity, Isaac Barrow's belief was entirely orthodox; but to say only that he was orthodox is to miss the distinctive flavor of his religious writings. John Wallis was orthodox, and Barrow contrasted sharply with Wallis. With Wallis orthodoxy was too often the lifeless repetition of received formulae. With Barrow it was agreement with tried and accepted solutions to spiritual problems deeply felt and studiously examined. Barrow was too concerned with the problems to become absorbed in the formulae; and when the formulae came from his pen suffused with the insight of his personal experience, they seemed like new truths freshly perceived, however orthodox and traditional they might be. Barrow was labeled, and to a degree even stigmatized, with a name; he was called "the exhaustive preacher." Charles II claimed that he was unfair because he left nothing for others to say on his topics. More than a model of expository style, however, he was a man of deep spiritual insight who found satisfaction for his most intimate aspirations in the Christian religion.

The irreducible ultimate at the heart of Barrow's religious thought was the nexus of love binding man to God—and God to man. Around this core he constructed his discourses and sermons. With other virtuosi Barrow found matter in the works of God that provoked awe for the Creator's wisdom and reverence for His power, but more keenly than either of these emotions he felt love for the Father's goodness. All things, even adversity (sent as he knew to correct him) inspired his love of God, so that Barrow's religious experience transcended the level of rational definition and demonstration. Love of God, as he conceived of it, is an act of the heart and the will detaching their affections from earthly things and returning to the source of all felicity. Although intellectual aids such as recognition of the Lord's excellence may promote love of God, in the end it must be the gift of the spirit, gained through personal communion with the Father. As other friendships are nourished by familiar converse, "so even by that acquaintance, as it were, with God, which devotion begets, by experience therein how sweet and good He is, this affection is produced and strengthened. As want of intercourse weakens and dissolves friendship, so if we seldom come at God, or little converse with Him, it is not only a sign but will be a cause of estrangement and disaffection toward Him." In prayer we not only behold God's excellencies but in a manner feel and enjoy them; our hearts, being warmed and softened by desire, become more susceptive of love. "We do, in the performance of this duty, approach nearer to God and consequently God draws nearer to us . . . and thereby we partake more fully and strongly of His gracious influences; therein indeed He most freely communicates His grace; therein He makes us most sensible of His love to us, and thereby disposes us to love Him again."[14] In reply to the yearning of the supplicant's soul, God's love descends to fill his bosom, to fill it until love transfigures his whole being, pervading his every act and remaking his life in the image of God.

One interpreter of Barrow has said that he devoted his sermons to morality. In a sense the statement is correct, but what a different morality it was from the stodgy respectability expounded in the volumes of natural religion. Barrow preached good works infused with charity, morality inspired by love. The spirit of the act was all important to him, and not the act itself. Believing that such morality is the necessary product of true religious fervor, he declared that Christianity without good works is dead. That is to say, the love of God must pervade a man's conduct and behavior and make of him a new person. Charity, then, the root of religion, is also the root of morality:

> We may consider that charity is the best, the most assured, the most easy and expedite way or instrument of performing

14. Ibid., 2, 287–8.

> all other duties toward our neighbor. If we would despatch, love, and all is done; if we would be perfect in obedience, love, and we shall not fail in any point; for "love is the fulfilling of the law, love is the bond of perfectness." . . . Charity giveth worth, form, and life to virtue, so that without it no action is valuable in itself or acceptable to God. Sever it from courage; what is that but the boldness or fierceness of a beast? from meekness; and what is that but the softness of a woman or weakness of a child? from courtesy; and what is that but affectation or artifice? from justice; what is that but humor or policy? from wisdom; what is that but craft and subtlety? What means faith without it but dry opinion; what hope but blind presumption; what alms-doing but ambitious ostentation; what undergoing martyrdom but stiffness or sturdiness of resolution; what is devotion but glozing or mocking with God; what is any practice, how specious soever in appearance, or materially good, but an issue to self-conceit or self-will or servile fear or mercenary design? . . . But charity sanctifies every action and impregnates all our practice with a savor of goodness, turning all we do into virtue.[15]

Love is the basis of both tablets of the law; Barrow believed that they may both be summed up in the injunctions to love the Lord thy God with all thy heart and to love thy neighbor as thyself.

In Christian charity man finds his true being; in its practice his soul grows and develops to reach its fulfillment. Barrow suggested that love of others is the truest form of self-love. Narrow self-love, selfishness in the usual sense of the term, shrivels and constricts the soul, withdrawing it from all pure delights; it distorts and deforms the soul, destroying its inherent beauty. Without charity all of the best faculties of human nature become vain and fruitless, even noxious and baneful to us. What is reason worth, Barrow asked, if it serves only to plot sorry designs or to transact petty affairs? What good is wit if it be spent only in making sport or hatching mischief? What is the purpose of knowledge if it be not applied to instruct, direct, and console others? What does wealth matter if it be hoarded uselessly or flung away

15. Ibid., pp. 395–8.

in wanton profusion? What is virtue itself if it be buried in obscurity yielding no benefit to others? When our gifts are infused with charity, however, they become precious and excellent. They are great in proportion to the greatness of their use. Uncharitableness, therefore, should be shunned and loathed as that which robs a man of all of his ornaments and advantages, "for without charity a man can have no goods but goods worldly and temporal; and such goods thence do prove impertinent baubles, burdensome encumbrances, dangerous snares, baneful poisons to him." In charity, on the other hand, we reach out and grasp the whole world to ourselves; we build on to ourselves from the rest of creation. Charity endows us with all the world has that is precious and fair. Our neighbor's wealth enriches us if we feel content in his possessing it. His preferment advances us if our spirits rise with it. His pleasure delights us if our hearts triumph in it. "This is the divine magic of charity, which conveys all things into our hands and instates us in a dominion of them whereof nothing can disseize us."[16] Isaac Barrow preached the perfection of the human soul, its growth, blooming, and fruition when irrigated with divine love. In the exercise of charity man not only fulfills the law; his own being approaches the divine image. Salvation, then, is not an extraneous matter, wages paid for good works performed; heaven is but the continuation and completion of a process well begun in this life.

The doctrines concerning Christ had obvious meaning in the context of Barrow's religious thought. On the one hand Christ came to pay man's debt to God; on the other He came to restore fallen man to his original image. In Christ human nature divests itself of sin and realizes its full potential. His life was the model of conduct irradiated with love. Through Him all men find their individual fulfillment. Since man in his fallen condition is unable to cast off all the concerns of the flesh and to live and grow in the realm of the spirit, Christ sends the Holy Ghost to assist man's efforts and to regenerate his soul. Jesus is the personification of divine love, the conduit as it were between God and man. Through His mediation man may drink from the fountain of life and be restored to health. Christianity was more than a set of

16. Ibid., pp. 388–91.

received dogmas to Barrow. While the exponents of natural religion nodded assent to the belief in Christ but focused their attention on other questions, Barrow centered his religion on the figure of the Redeemer. When he expressed his belief in Christ, he was reaffirming the core of his whole religion.

The temper of Barrow's religion can be felt in the contrast between his sermons on morality and the ethical teachings of natural religion. John Wilkins' *Natural Religion* and John Ray's *Persuasive to a Holy Life* tried to prove that Christianity leads to worldly happiness as well as to future salvation. While they did not interpret happiness as material prosperity, they did tend to equate Christian ethics with bourgeois values and to reconcile Christianity with the world. To Wilkins and Ray, Barrow would have replied that the Christian life is, as they contended, the life of happiness—but not happiness defined in the world's terms. Material wealth and sensual pleasure are the values of the world, and the world pursues them despite the fact that they do not produce enduring satisfaction. The Christian, too, must be attracted by them, and to reject the values and pleasures of the world is a work more difficult than the simple choice of rational self-interest pictured by Wilkins and Ray. To fulfill the demands of Christian living a man must steel himself for a never-ending struggle with the flesh and an unremitting effort of self-discipline. He must forego desires that appear natural to him; he must undertake duties that at first can seem only burdensome. Barrow did not doubt that Christianity is at war with the world. Not only is Christian virtue unlikely to beget material wealth; it is more apt to bring a dearth of worldly riches. Christian morality, as he thought of it, is a life of struggle and toil—struggle and toil commensurate with the goal at which it aims.

Barrow's religion can be found in his sermons, which rank as masterpieces of pulpit artistry in an age of great preaching. His religion can also be found in his life, and his conduct set the mark of sincerity on his spoken word. The charity that he preached shone in his character. As far as records reveal, he left no enemies behind him, and his contemporaries spoke of him only with love and respect. Many anecdotes about him have survived, showing how unconcerned he was with the normal life of the world. The length of his sermons wearied the unregenerate Philistines to whom he preached, and they resorted to various stratagems to shorten them. The vergers in Westminster Abbey once cut him off by having the organs played "till they had blowed him down." His uncouth appearance—for his neglect of his person went to the extreme—affronted the respectability of his day. Many of the anecdotes about Barrow group around these two themes: kindly stories, really, about a beloved and revered figure—dear, eccentric Uncle Toby. Isaac Barrow was so deeply absorbed in spiritual matters that he lived in a world apart. The vital reality of Christianity within his soul explains why he could not rest satisfied with the verbal propositions of natural religion.

John Mapletoft (1631–1721), a successful doctor, reflected the pattern of Barrow's life both in his career and in his convictions. In spite of his success as a doctor, Mapletoft found that medicine could not fulfill all of his aspirations. Why he entered the profession in the first place is uncertain. Ward states that he took up medicine because he found the ministry closed to one of Anglican profession when he completed his B.A. during the interregnum. Since he did not begin to study medicine until 1660, the argument is not convincing. Whatever his original motives were, John Mapletoft was a minister of God who found himself practicing medicine by mistake. Since he was a sincere man, Mapletoft served his profession loyally. Although he did not have the genius to make a major contribution to science, as Barrow did when he occupied the Lucasian chair, Mapletoft did achieve some eminence, rising to the Gresham chair of physic. For a time he participated actively in the Royal Society, of which he became an officer. He followed the progress of medical science with intelligent interest and applauded the work of Thomas Sydenham, whose writings he translated into Latin for publication. Nevertheless, he was uneasy as a doctor; his conscience could never rest comfortably under the responsibility for his patients' lives. His religious interests led him to choose his closest friends from among the clergy. In 1679, when he was nearly fifty years old, Mapletoft decided to abandon medicine and enter the ministry. After being confirmed in 1683, he took up a living in Northamptonshire; and his satisfaction in his new calling was such that he never regretted the change. Like Barrow, Mapletoft felt the call of spiritual duties so strongly that he abandoned a successful career to devote himself to them entirely.

Mapletoft's religious writings—a couple of sermons that were printed, a moderately long exposition of the *Principles and Duties of the Christian Religion* (1712), a more modest handbook of aphorisms named *Wisdom from Above* (1714), a collection of prayers, and a collection of moral proverbs—reflected the type of religious convictions shown by his change of profession. The opening pages of the *Principles and Duties of the Christian Religion* repeated the demonstrations of natural religion, but the primary interest of the book and of his other works centered on the spiritual relationship between God and His creature, man. Where the exponents of natural religion saw in God chiefly omnipotent power which had created the universe, Mapletoft thought of Him as goodness and love radiating through all of His works, a mighty spiritual force operating in the heart of man. Man was created in the image of God; and though the fall has dimmed his faculties, there is still a spark within him than can respond to the divine spirit. In religious exercises, in the employment of the understanding to know God and of the will to love Him, a man brings his highest faculties into play; directing them toward their proper object, he realizes the full potentiality of his nature. Attaching his affections to God, directing his every effort to obey the divine will, subordinating all else to the supreme goal, a man rises above the petty concerns of the world into the realm of true and enduring values. God's love in turn descends into the responsive soul, fertilizes its growth, and helps it to spread and ascend; as it perfects itself, the soul prepares itself for the everlasting communion with God to be attained finally in the life hereafter. "The great mystery of Christianity," Mapletoft wrote, "is the perfecting of the human nature by the participation of the divine nature." [17]

Mapletoft demanded not external conformity to the moral law but internal and spiritual assent and aspiration. True Christian

17. John Mapletoft, *Wisdom from Above: or, Considerations Tending to Explain, Establish, and Promote the Christian Life, or That Holiness without Which No Man Shall See the Lord* (London, 1714), Pt. I, p. 51.

faith must be an operative influence in a man's life, reforming and transforming his soul. When a man has liberated himself from the bonds of sin to enjoy the liberty of pure obedience, he reaches the culmination of his humanity. Then is a heaven created within his own bosom; love, peace, tranquillity, contentment, the spiritual virtues of Christianity—all hold him secure above the strife of worldly affairs. Flowing through his whole being, love pervades his conduct, spiritualizes his relations with other men, frees him from the resentments and quarrels of the natural man. Man saved, man transformed and spiritualized, is man exalted and ennobled, perfected in the divine image that he first wore. "All love is transforming," Mapletoft declared; "the love of God to man transformed Him into the likeness of man, and the love of man to God transforms men into the likeness of God." [18] On the other hand the man mired in sin, in sensuality, avarice, pride, and all of the other snares of the world, gradually darkens the light within and destroys his own humanity. Before his eternal punishment he builds a hell in his own breast which could have sheltered a heaven of bliss.

John Mapletoft was incapable of reaching the heights that Barrow scaled. His *Principles and Duties of the Christian Religion* was not a masterpiece of Christian apologetics. Nevertheless his purpose, however hesitantly executed, did resemble the goal at which Barrow aimed. Mapletoft tried to present religion as more than a set of rationally demonstrable propositions. Christianity to him was a response to the daily experiences and needs of the individual man. It must first of all be true, of course; it could not be simply an imaginative dream to console human frailty. To help demonstrate the truth of Christianity Mapletoft was happy to make use of natural religion. Proving the rational validity of religion was only the preliminary step, however. Christianity, Mapletoft thought, is the solution to human shortcomings. True faith promotes man's spiritual growth; it enables him to cast off the fetters of sin, to ennoble and perfect his character, and to win eternal salvation. A spirituality not unworthy of the Christian religion illuminated Mapletoft's writings. His attention was riveted on problems that went beyond natural religion. To min-

18. Ibid., pp. 92–3.

ister to those problems he abandoned a flourishing medical practice to become a preacher of Christian salvation.

The religious beliefs of the virtuosi who wrote on natural religion did not diverge significantly, except in a few cases, from the beliefs of Sir Thomas Browne, Isaac Barrow, and John Mapletoft. Both Barrow and Mapletoft accepted and used natural religion; and although Browne was too singular in his thought to employ its demonstrations, he did share the conclusion that religion can be supported by reason. Nearly all of the exponents of natural religion, for their part, believed in the revealed doctrines of Christianity. The great difference between the religious works of Browne, Barrow, and Mapletoft and the discourses on natural religion arose from the questions that they examined. Those absorbed in natural religion devoted their books to the proof of God's existence and of His governance of the world, while the other three men were concerned primarily with the development and salvation of the human soul—that is, with the spiritual teachings of Christianity. Although the virtuosi who wrote on natural religion continued to believe in the revealed doctrines of Christianity, they gave them very little attention. Without denying that some doctrines of Christianity transcend reason, they emphasized only what was rationally demonstrable. They pushed to the side and ignored the concept of redemption and the traditional doctrines of Christianity grouped around it. They argued for a religion which was little more than a rational moral code. Whereas the specifically Christian elements could not have been removed from the religious thought of Browne, Barrow, and Mapletoft without destroying the whole, the natural religion that dominated the minds of other virtuosi was not insolubly connected with Christianity; and taken by itself their natural religion was indistinguishable from deism.

CHAPTER 7

Reason and Faith

Human reason is very much subject to mistake about divine things when she presumes to pronounce dogmatically of them upon her own stock of knowledge . . .

Boyle, *Four Conferences about as Many Grand Objections against the Christian Faith*

In all things . . . where we have clear evidence from our ideas and those principles of knowledge I have above mentioned, reason is the proper judge; and revelation, though it may in consenting with it confirm its dictates, yet cannot in such cases invalidate its decrees.

Locke, *Essay Concerning Human Understanding*

SINCE most of the virtuosi made the exposition of natural religion their characteristic religious interest, they were directly concerned with the competence of reason in religion. Anyone who accepted and employed the demonstrations of natural religion acknowledged in effect that reason can perceive a considerable amount of truth in religion; and even those who did not much care for natural religion were willing, in accordance with a well established tradition in Christianity, to accept reason as the complement of revelation. With the growing importance of natural religion the competence of reason in religion had to be re-examined to determine its exact scope and nature. Can reason achieve absolute certainty in religious matters? What is the relation between reason and faith? Can man, who is endowed with the gift of reason, properly accept articles of belief above reason? Some answers to the questions inevitably had to be given by the virtuosi who investigated natural religion. Mostly the replies were stated implicitly rather than explicitly. In the attitudes they took toward natural religion and toward the Bible the virtuosi were adopting definite conclusions on the dividing line between reason and faith, even though they may not have stopped to analyze their position. Four of them did specifically consider the relations of reason and religion. John Wilkins, Robert Boyle, Joseph Glanvill, and John Locke examined various aspects of the problem and advanced definite theories.

Since the four virtuosi did not agree on the meaning of "reason," it is difficult to compare their conclusions. Whereas Glanvill considered that the faculty of reason carries the innate impression of certain divine truths, Locke argued that it knows only those empirical facts which sensation and reflection print on the mind, together with the conclusions drawn from them. Fundamentally, however, the virtuosi were discussing the competence of the human intellect more than the nature of reason; and the differences between their concepts of reason are less important than the comparison of the roles they assigned to revelation. The ultimate question was the relative authority and importance—importance as much as authority—of unaided human faculties and supernatural revelation in discovering religious truth. All of them agreed that the human faculty of reason is God's creation and that the material on which it works—whether innate ideas or external nature perceived through the senses—conforms to the reason of the Almighty. The issue was the degree to which supernatural knowledge supplements and corrects reason.

When John Wilkins first approached the problem in the *Discourse Concerning a New Planet* (1640), he was not really discussing reason in religion. At that time his purpose was to defend the right of scientific inquiry, but his argument indicated where he thought the boundary between reason and revelation lies. Truth, he suggested, consists of two parts, philosophical and theological. The first is to be found by rational inquiry; the second is revealed in the Scriptures. To seek philosophical truth in the Bible would be to confuse the boundaries of the two forms of truth. While the Scriptures are meant to guide us in matters of faith and obedience, God did not intend to settle philosophical problems by revelation. He left the quest for philosophy as a labor to occupy man's leisure, thus to divert him from sin. Since Wilkins' intention in the *Discourse* was to answer the objections of religious orthodoxy to Copernican astronomy, he did not consider the competence of rational inquiry in religion. He did affirm the superior authority of reason, in this instance equated with natural philosophy, over revelation outside the boundaries of theology. In asserting that astronomy gives positive aid to religion by proving the existence of God, he opened the way to natural religion, although, of course, he did not challenge revelation as a source of religious truth in any way.

In the *Principles and Duties of Natural Religion* (1675) Wilkins returned to the same division between philosophy and religion, viewing it this time from the side of religion. His purpose was twofold: to show that reason is not invalid in religion, and at the same time to preserve religion from excessive demands for rational demonstration. In essence, his position repeated the stand taken in the *Discourse* over thirty years before. Although reason is able to reach a high degree of probability—amounting nearly to certainty—in some religious matters, religion and philosophy are nevertheless distinct fields. Philosophical demonstrations of religious propositions are not to be expected. Wilkins began by distinguishing the different kinds of evidence by which knowledge is gained—virtually adopting empiricism but accepting the validity of mathematics as well—and the different degrees of assent. When assent admits no reasonable doubt, it is called knowledge or certainty. He divided certainty itself into different degrees. Physical certainty, which depends upon the direct evidence of the senses, and mathematical certainty which depends upon apriorisms, can be called infallible certainty, but below them lies the plane of moral certainty, assent to things lacking absolute proof. Wilkins called the latter form of certainty indubitable certainty, a loaded phrase by which he meant that a reasonable man can have no reasonable doubt of the certainty of the thing in question, although the possibility of error exists. Finally, when evidence is less plain and clear, assent is a matter of probability or opinion. All truths are equal in themselves, he continued, but all cannot exhibit the same degree of evidence to us. Thus the existence of some historical person cannot be proved by the direct testimony of the senses, and yet his former existence is a truth equal to one directly observed by our senses. Nothing can be more irrational than to doubt or to deny the truth of anything because it cannot be demonstrated by a kind of evidence of which it is incapable. "A man may as well deny there is any such thing as light or color because he cannot hear it, or sound because he cannot see it, as to deny the truths of other things because they cannot be made out by sensitive and demonstrative proofs." [1] Different kinds of things require different sorts of proof; demonstrations from sensible evidence must not be expected for everything. "When a thing is capable of good proof in any kind," Wilkins said, "men ought to rest satisfied in the best evidence for it which that kind of things will bear and beyond which better could not be expected supposing it were true." [2]

As he insisted in the *Discourse* that philosophy and religion are distinct fields of knowledge with their own sources of truth, so Wilkins maintained in *Natural Religion* that religion must not be expected to advance demonstrations proper to philosophy. Confirmed physical facts are absolute knowledge, and mathematical conclusions, which belong to a self-dependent system, demand complete assent. Not everything attains the same degree of certainty, he contended, but man cannot live by physical science and mathematics alone. Many things of a practical nature can be known only with a lower degree of certainty, yet in matters of supreme importance men cannot withhold assent even when the possibility of error exists. Wilkins believed that in order to live wisely and well men must be content with moral certainties or even probabilities that are judiciously weighed. Skeptics in religion refuse assent, not because the weight of evidence is against religion, but merely because the demonstrations are imperfect enough to allow them to raise doubts. " 'Tis sufficient," Wilkins concluded, "that matters of faith and religion be propounded in such a way as to render them highly credible, so as an honest and teachable man may willingly and safely assent to them and according to the rules of prudence be justified in so doing." [3] It would be unwise, he continued, for demonstrations of religion to be necessarily certain; for then no room would be left for faith. To choose faith when it is not absolutely certain is a positive virtue.

1. Wilkins, *Natural Religion,* pp. 22–3.
2. Ibid., p. 25.
3. Ibid., p. 30.

In the main body of the work Wilkins applied his principles to the demonstration of natural religion. Fundamental to the argument of the whole book was the proof of the existence of God. From God's existence Wilkins deduced His attributes and finally the religious duties of man. The strength of the argument depended on the certainty of the existence of God. Wilkins offered four demonstrations—the universal consent of men in all times and places, the fact that the world is not eternal, the skillful contrivance of all natural things, and the works of providence in the government of the world. In conclusion, he agreed that his proof of the existence of God is not infallibly certain, but in so important a matter men must accept the best demonstration that the case admits. Immediate sensible evidence of the existence of an invisible being is impossible. Certainly the evidence is heavily in favor of God's existence. Even if the evidence were only balanced, Wilkins decided finally, men should accept the conclusion that most favors their own self-interest. If God does exist, eternal damnation will be the price of disbelief; the prudent man will choose to avoid such a danger.

The ultimate implication of Wilkins' principles imposed rather strict limitations on human faculties. True, he reached the conclusion for which he set out, demonstrating to his own satisfaction that human reason can find sufficient proof of God's existence to make it a reasonable belief. In contending that it was only sufficient proof—in his own terms moral certainty rather than absolute certainty—he was striking not at the validity of religion, which was unquestioned in his mind, but at the competence of human reason. In Wilkins' usage "reason" was practically equated with the method of natural science. Those things which the unaided reason can know with absolute certainty are truths learned from the evidence of the senses or from mathematical propositions, while anything advanced on less secure evidence can at best reach moral certainty. In his proof of the existence of God he introduced one exception to his customary usage of "reason"; in using the universal consent of all men as evidence of God's existence, he implied that the idea of God is innate. For the most part, however, Wilkins compared the certainty of natural religion to the certainty of natural philosophy. When he decided that natural religion is less certain, his conclusion was that natural philosophy —or human reason—is an instrument of limited value. All truths are equal, although man is incapable of knowing them equally well. Natural philosophy and religion are two different things. Faith is still faith. Although Wilkins did not take up the authority of revelation in his study of *Natural Religion,* his principles left no doubt that he placed it on a higher plane than human reason —as a source of religious truth.

Nevertheless a disquieting note crept into his book. Too often he used his argument more as a cloak to shield weaknesses than as a lamp to illuminate truth. First he concluded that the principles of natural religion are morally certain, that is, in his words, indubitably certain, though not infallibly so. Mixing the degrees of assent which he had set up, he later asserted that the principles of natural religion are as certain as material phenomena, which he had called infallibly certain. He contended that it is not absolutely certain that the sun will appear in the morning or that a house will continue to stand from one moment to the next. Finally, he said that even if religious principles are only probable, enlightened self-interest demands that they be accepted. He defended uncertainty on the grounds that faith is a laudable virtue; to give faith scope God has deliberately left things obscure. By constantly shifting his ground, Wilkins weakened his position immeasurably and left the impression that he was arguing from desperation, uncertain himself of the strength of his own case. Although none of the virtuosi pursued the doubts that Wilkins implied and challenged the principles of natural religion, his suggestion that even natural religion could not stand the full test of reason was an ominous warning of the conclusion to which reason might lead if it were rigorously and fully applied.

Robert Boyle extended Wilkins' examination of the competence of reason in religion. Where Wilkins was concerned to defend natural religion from the demand for infallible demonstrations, Boyle's inquiry into reason and religion devoted itself primarily to the vindication of revelation and things above reason. Boyle's ready employment of natural religion to prove the existence of God demonstrates sufficiently that he was not an obscurantist who placed all religious truths beyond the scope of rational investigation. He maintained that religion should be thoroughly examined; it is too important to be accepted on trust. In the *Excellency of Theology* (1674) he asserted that anything not susceptible of a cogent and rational proof cannot be an essential of Christianity. It is unreasonable to think that God, Who sent His only Son into the world to promulgate the true religion and worked miracles to attest to it, "should not propose those truths, which He in so wonderful and so solemn a manner recommended, with at least so much clearness as that studious and well-disposed readers may certainly understand such as are necessary to believe."[4] When he said that doctrines necessary to salvation are comprehensible to the human intellect, Boyle did not mean to imply that all of religion falls within its competence. Much as he believed in the function of natural religion, he was firmly convinced that Christianity embraces doctrines extending well beyond the limits of human understanding. In one of his unpublished papers on the causes and remedies of atheism Boyle discussed the objection to Christianity, that the notion of God and His attributes is too obscure to be accepted by reason. "I cannot allow," he replied, ". . . that the intellect of man is the genuine standard of truth, so that whatever surpasses his comprehension must not be admitted to be."[5] Those who think that their own reason enables them to soar into heaven and to discover the most sublime truths of religion without the assistance of revelation are like flying fish, whose wings, or fins, do indeed raise them out of the water but cannot carry them to any great height or long sustain them. The proper function of human reason is not to teach us supernatural things but to lead us to a supernatural teacher—that is, to God—and to defend the things that He teaches from the charge of being contradictory or impossible.[6] The mind of man is so made that it can recognize its own want of light for some purposes and the insufficiency of the best ideas that it can frame about some things. It can discern the imperfection of its own faculties enough to realize that some objects are disproportionate

4. Boyle, *Works*, *4*, 41.
5. Royal Society, Boyle Papers, *6*, fol. 330.
6. Boyle, *Works*, *6*, 714–15.

to its powers. Thus reason itself, in Boyle's opinion, prepares the acceptance of doctrines above reason.

In a *Discourse of Things above Reason*, published in 1681, Boyle analyzed the nature of things beyond human comprehension. He separated them into three categories. There are, first, those things the nature of which we cannot comprehend. All of the intellectual beings of a higher order than the human soul, including God of course, belong to the first group. In the second group are things with properties which we cannot understand. Although their existence cannot be denied, we are unable to conceive how they can be as they are. How, for instance, can the diagonal of a square be incommensurable with the side? Finally, in the third category, some things are above reason because they seem to contradict acknowledged truths, as free will appears to be incompatible with divine foreknowledge of human actions.[7] Although he did not mention them in this passage, Boyle usually included religious doctrines that reason can understand, but not discover among the things above reason. Thus the essentials of salvation are comprehensible once God has revealed them. In order to argue the credibility of religious doctrines above reason, Boyle liked to point out that aspects of natural philosophy are impenetrable by the human intellect. Man is unable to know the true nature of bodies, for instance, or to comprehend how an immaterial soul can move a body. The examples from natural philosophy were mentioned only in passing; his analysis was primarily concerned with religious doctrines above reason. Boyle tried to maintain a moderate position distinguished from either of two extremes. While, on the one hand, he accepted doctrines beyond human comprehension, he also believed in reason's sufficiency within its proper limits; and he wished to close the door firmly against impostures. He took severe exception to Tertullian's dictum, *Credo quia impossibile est.* True, we should and do accept on faith doctrines beyond our understanding, but we do so in spite of the difficulty, not because of it. Tertullian's formula would open the way to every wild obscurity.[8] We must test every

7. Ibid., *4*, 407.
8. Royal Society, Boyle Papers, *1*, fol. 70.

belief. If we assent to things that we do not fully understand, we do so only when we have sufficient cause for assent. Articles of belief must come with evidence that they are divine and therefore demand our acceptance. The effect of Boyle's conditions was to limit things above reason to the biblical revelation. He would have nothing to do with latter-day revelations, the "chaos of enthusiastic notions and dictates," as he called them.[9] The Holy Scriptures, however, are supported by miracles which prove them to be the word of God. Truths above human comprehension, such as the attributes of God, and truths beyond human discovery, such as the future state of man, must be accepted and believed because the Bible reveals them.

Boyle devoted some of his most penetrating thought to the examination of things above reason, "privileged things" in his terminology, and to the consideration of their authority in relation to rational philosophy. In addition to *A Discourse of Things above Reason* important passages in the *Christian Virtuoso* and a large number of fragments, which were intended as parts of a treatise named *Four Conferences about As Many Grand Objections against the Christian Faith,* deal with "privileged things." Boyle was concerned to defend both reason and revelation—to show on the one hand that revelation does not destroy the use of reason, and on the other hand that revelation is not obliged to conform to human philosophy. The second purpose occupied most of his attention. In the *Christian Virtuoso* Boyle introduced a basic proposition. A man should accept things taught by experience that without experience would be judged impossible. Boyle argued that many things were being discovered in natural science that were formerly thought to be false. Before the 16th century no one believed that bodies of unequal weight accelerate equally when falling freely. Before explosives were discovered, no one believed that a little black powder could knock down a wall. Actual experience, then, is of higher authority than mere theories, however conclusive their demonstrations may appear. Personal experience, which a man acquires by his own sensations, is not the only valid experience. Historical experience, conveyed by the testimony of some other man to whom it was personal experience,

9. Ibid., *4,* fol. 66.

carries equal authority; and so does theological experience, which God reveals either directly, as He did to Job and Moses, or indirectly through prophets, inspired persons, and inspired Scriptures to men at large. If we accept facts that seem to contradict reason in natural philosophy, we should do so more willingly in spheres where our understanding is incompetent to judge.[10] Boyle was basing his argument on two principles which underlay his whole analysis of "privileged things": first, that human philosophy is imperfect and is constantly forced to revise itself as experience uncovers new facts; second, that human philosophy is unable to judge finally on certain matters. Both principles argue that man is not justified in rejecting revelation solely because it contradicts his philosophy.

The *Discourse of Things above Reason* presented six rules to guide inquiries into "privileged things." The first rule, that we admit no affirmations about "privileged things" without sufficient evidence to substantiate them, repeated his caution about impostures. The second rule was meant to balance the first: we should not be too forward in admitting negative propositions or in rejecting positive ones. If some truth about "privileged things" be proved by competent arguments, the third rule stated, we ought not to deny it merely because we cannot explain or perhaps even conceive the means by which it can be so. For instance, we are unable to comprehend how an immaterial substance like God can move matter. Fourth, when we treat of things above reason, we are not bound to think that everything seemingly contradictory to some received dictate of reason is false. Fifth, we do not have to reject a "privileged thing" because it seems inconsistent with some other thing that we believe to be true. Sixth, in "privileged things" we should not always condemn an opinion merely because it entails inconvenient or seemingly absurd consequences.[11] Throughout Boyle's argument it is clear that "reason" in his normal usage corresponded very closely to natural philosophy. He maintained that human philosophy has been drawn from the material finite world; we have no grounds to believe that it is universally valid, that it can be applied to the

10. Boyle, *Works, 5,* 525–9.
11. Ibid., *4,* 449 f.

realm of the eternal, infinite, and spiritual. Moreover, man is a finite being with a limited intellect. Contradictions in his eyes may be resolved into harmony by a higher intellect. In one of the fragments from the *Four Conferences* Boyle gave a concrete example of the limitations imposed on philosophy by the subject from which it is drawn. The discovery of the Orient and America so increased information about the phenomena of nature and disclosed such a wide variety of her productions unknown to the Greeks and Romans that extensive changes in natural philosophy necessarily followed.[12] If confinement to the phenomena of Europe and the Mediterranean basin could distort natural philosophy, how much more must confinement to material objects distort human judgments about spiritual questions.

In the *Christian Virtuoso* Boyle carried his analysis further, trying to show that apparent contradictions between reason and revelation are due to the limitations of the human intellect. There is a distinction, he thought, between human philosophy and right reason. "Reason" can signify several things. Sometimes it means the superior faculty of the mind furnished with its own notions and axioms. "Reason" may also mean an "organical" thing, not the faculty alone but the faculty managing a system of ideas and propositions furnished by the sciences and arts. Although such a system is a tool which reason makes for itself, once it is made, it regulates the operation of reason. "Reason" in the second sense is called natural reason or philosophy. This was Boyle's normal use of the word. There is a third use of "reason" in which it is also considered as acting organically, in the sense already given, but with a more noble instrument, revelation. Reason in itself is the same in speculations of all kinds. Acting organically through an instrument, reason is more limited in respect to some objects than to others. It needs special sets of notions to deal with different objects. Right reason is reason fully informed, or at least as fully informed as is necessary to pass a sound judgment on the thing in question. Right reason, then, has a broader extent than philosophy, which is reason informed only by natural light. Reason informed by natural light is not competent to refute reason enlightened by revelation. This was to repeat his conclusion that

12. Royal Society, Boyle Papers, *1*, fol. 118b.

philosophy based on the material world cannot validly judge the realm of the spiritual and the eternal.

The objection may be raised that truth is truth, and that two truths cannot contradict each other whatever the word "reason" may signify. In reply Boyle declared that all supposed truths do not have the same certainty. A distinction must be made between probationary and absolute truths. While a probationary truth appears valid in the light of available information, it must stand ready for revision in the face of new discoveries. Absolute truths, on the other hand, are the theoretical principles and axioms that provide the foundation for all reasoning. The proposition that two quantities equal to a third quantity are equal to each other, and the proposition that a thing must either be or not be are truths of the latter order. They are absolute and eternal. They are not subject to contingent circumstances. There has never been a time when they were not accepted. Here Boyle was introducing a different concept of reason, universal rules of divine reason that both subsist apart from man and are impressed congenitally upon his intellect. While revelation may contradict a probationary truth, which is a product of human reason, it obviously cannot contradict an absolute truth which is the product of divine reason. Since absolute truths are the principles upon which assent to revelation itself must be grounded, any revelation that seems to contradict them must be interpreted in a way that removes the difficulty.[13] "If revelation makes us refuse the authority of philosophy," Boyle asserted, "'tis in such points where reason itself tells us that philosophy ought to have no authority, unless about such points revelation be silent. And where revelation contradicts but the vulgar doctrines of the School, philosophers themselves ought, as well as divines, to condemn such philosophy."[14] Boyle insisted on the distinctions between true reason and different schools of philosophy, and between true revelation and different sects of theologians. He refused to be burdened with the task of reconciling an Epicurean philosopher with a peripatetic divine. Between true reason and true revelation, however, Boyle was certain that no contradiction is possible.

13. Boyle, *Works, 6,* 707–13.
14. Royal Society, Boyle Papers, *1,* fol. 37.

Far from contradicting each other, reason and revelation reinforce and supplement each other in Boyle's opinion. Revelation supplies truths that man can never attain by himself—the order and time of the creation, and the particulars of Christian theology such as the Trinity, the incarnation, and the redemption. Revelation also reveals with certainty things that natural reason can perceive only dimly, such as God's providence and the immortality of the soul. Through revelation man is freed from the many errors and superstitions that he falls into when left to his own powers. Revelation serves human reason as the telescope serves the eye; it both discovers new things and reveals more clearly things faintly visible without it.[15] If the truths proposed by revelation are a burden to reason, they are such a burden as feathers are to a hawk—instead of hindering its flight, they enable it to soar.[16] Boyle maintained that everything disclosed by revelation is fully consonant with true reason, that the two together form one organic body of knowledge. "Right reason and divine revelation being both of them emanations from the Father of lights, there is no likelihood that they should contradict one another." [17]

Robert Boyle, like John Wilkins, was defending religion from the demand for demonstrations like those proper to natural philosophy. Although he readily admitted and even aggressively claimed that man is able to demonstrate the propositions of natural religion, he insisted that only the word of God can reveal the superior truths of Christianity. While human reason can and must examine the credentials of revelation, it is simply incompetent to judge the revelations themselves. In effect Boyle was distinguishing between pure reason, thought of as natural law, and the human faculty of reason. To the mind of God all things are rational. Man on the other hand is both limited and prone to err. In those matters about which God has spoken man should recognize his incapacity and submit to authority, for human reason is not the measure of the universe.

With Joseph Glanvill and John Locke the purpose of analyzing the role of human reason in religion was entirely different. Where

15. Ibid., *4*, fols. 10 f.
16. Boyle, *Works*, *4*, 16.
17. Royal Society, Boyle Papers, *1*, fol. 86.

Wilkins and Boyle were concerned to prove that religion cannot be completely demonstrable and to expose the limits of the human intellect's authority in religion, both Glanvill and Locke were interested more in asserting the competence of reason than in challenging it. The keynote of Glanvill's thought was sounded in his first book, *The Vanity of Dogmatizing* (1661). More than an exercise in systematic skepticism, the *Vanity of Dogmatizing* was a condemnation of dogmatic Aristotelian philosophy on the one hand and dogmatic sectarianism on the other. Glanvill's censure of Aristotelianism was not immediately connected with the question of reason and religion; his antipathy for sectarians, or enthusiasts, a hearty repugnance which crept into nearly everything that he wrote, was entirely concerned with it. Glanvill hated and feared enthusiasm, enthusiasm which had flourished and spread and disrupted society during the years when he grew to maturity, "those sickly conceits and enthusiastic dreams and unsound doctrines," as he described enthusiasm bitterly, "that have poisoned our air, and infatuated the minds of men, and exposed religion to the scorn of infidels, and divided the Church, and disturbed the peace of mankind, and involved the nation in so much blood and so many ruins." [18] Like many of the men in his age Glanvill saw Christianity under dire attack. On the one hand the atheists were challenging religion in the name of mechanical materialism. The sainted pages of the Bible, holy though they were, could not refute their arguments. Philosophy alone could ward off the danger. Meanwhile religion's strength was being sapped by a cancerous growth within its own bosom; the application of reason was needed to save Christianity from itself. The hordes of madmen, the unreasoning sectarians, were plunging all religion into disrepute. Already they had brought forth a host of absurd opinions which no reasonable man could accept— "ungrounded credulity cried up for faith, and the more vigorous impressions of fancy, for the Spirit's motions." [19] The very multiplicity of faiths called all into doubt. The fury with which each de-

18. Joseph Glanvill, *A Seasonable Recommendation and Defence of Reason in the Affairs of Religion, against Infidelity, Skepticism, and Fanaticisms of All Sorts* (London, 1670), p. 1.

19. *Vanity of Dogmatizing*, pp. 104–5.

nounced the persuasions of the others set a sorry example of Christian charity. The zeal of the sectarians produced less spiritual warmth than dogmatic fire which threatened to burn down the temple of worship. Unwarranted neglect of reason in religion was the source of these misfortunes. Only the vigorous reassertion of reason could preserve Christianity from its manifold enemies.

In a little essay, "Reflections on Drollery and Atheism," attached to *A Blow at Modern Sadducism* (1668), Glanvill recounted a supposed conversation with an atheist whom he converted to religion by rational arguments. When he asked the man how he had come to atheism in the first place, Glanvill was given to understand that he had descended through all the levels of sectarianism, not stopping until he reached the sink of folly and madness, Quakerism; and from there he took the remaining short step into atheism. Certainly, Glanvill remarked, when anyone who places his religion in the opinions and beliefs of a sect considers the vast variety of sects, the confidence and pretensions of each to authority, and the doubtfulness of the things they hold certain, he will be in great danger of atheism.

> And except he fix . . . upon the few, plain, acknowledged essentials of belief and good life . . . 'tis a miracle if he ends not there at last. For he having established this, that religion consists in the way or form of some party or other, and then having successively deserted those sects that had most of his favor and affection, and so passed from one to another through all the steps of descent, when at length he is fallen out with the last, he has nothing else to fly to but contempt of all religion as a mere juggle and imposture.[20]

Enthusiasm was the deadly foe of both reason and religion in Glanvill's opinion. Lacking solid foundation of any sort, its visions and revelations were the products of overheated and melancholy imaginations which were mistaken for reality. Sober and reasonable men could only be repulsed by its extravagance, and religion more than disbelief was likely in the end to suffer the most damage from its zeal. If religion was not destroyed, society would be, as Englishmen had good cause to fear. The true purpose of

20. *A Blow at Modern Sadducism,* p. 182.

Christianity was being lost in the controversies between sects. Instead of a way of godly life Christianity was becoming a series of dogmas, a set of notions, dry bones over which the sects snarled while the nourishing food of divine life spoiled untouched. Reason, then, solid and substantial reason, reason instead of overwrought imagination, reason which recognizes what is truly important and admits the limits of its own certainty—reason, above all else, was what religion most needed.

Manifestly Glanvill's purpose in examining the use of reason in religion differed from Boyle's, and the meaning implied in his frequent reiteration of "reason" derived from his purpose. Boyle was defending biblical revelation from the demand that it conform to the conclusions of natural philosophy; his use of "reason" was roughly synonymous with natural philosophy. On the other hand Glanvill, in attacking enthusiasm, used "reason" as the antonym of the enthusiast's supposed "imagination" or "fancy." By "reason" he meant the faculty of the mind that weighs evidence judiciously and separates the spurious from the true. In this sense "reason" was nearly equivalent to common sense—in which he believed the enthusiasts to be lacking. Glanvill also used "reason" in a broader sense, which reflected his affinity with the Cambridge Platonists. "Reason" so used meant the principles of universal reason which the Creator implanted in the minds of men. The latter usage was also directed against enthusiasm. It implied that the fundamental and important religious truths were the innate principles of reason—fundamental and rational principles opposed to the unimportant and fanciful conceits of the enthusiasts.

The Vanity of Dogmatizing poured Glanvill's scorn on the pretentiousness of the sectarians in thinking that they knew final truth. He did not mean to imply that man cannot know any religious principles with certainty. Despite the seeming intent of the *Vanity of Dogmatizing* to deny all certainty Glanvill only challenged dogmatism in opinions that appeared wrong to him. In religion the wrong opinions were minor and obscure points which had no importance and could not be proved either by Scripture or by reason. He did not question the certainty of the doctrines that he took to be fundamental. In *A Seasonable Recom-*

mendation and Defense of Reason in Affairs of Religion against Infidelity (1670) Glanvill discussed the whole question of reason and religion and tried to determine the authority of human reason when it examines religion.

He contended first that reason befriends religion by proving some of its basic articles and by defending all of them. Since revelation supposes the existence of God and cannot prove it, reason must establish this fundamental belief. Reason again is the sole judge of the miracles on which the authority of the Scriptures depends. Men therefore must rely on their reason to be certain of the fundamentals of Christianity. Glanvill maintained further that reason serves religion by defending the doctrines of faith revealed in the Bible, although it can neither discover nor demonstrate them. It can show how an article of faith is possible; or if that cannot be done, reason can prove that the article should be accepted anyway. If a man should assert that the episode of the ark is incredible, computation can demonstrate that the ark was big enough to hold the animals. If, on the other hand, someone should refuse to believe that Aaron's rod turned into a serpent, or that God and man were united in Christ, human reason cannot show the manner in which these things were accomplished. It can, however, argue that God can do things beyond our comprehension because His power is infinite. Since we accept natural things that are revealed by the senses even though we do not understand them, we ought to receive as well articles of faith revealed by God. Thus human reason supports religion even in those fields where it cannot itself judge.

Despite the obvious necessity of using reason, Glanvill continued, some people maintain that it is not to be admitted in religion. They claim that the Bible must be the sole authority, for God says that the wisdom of this world is foolishness. The fall so dimmed human faculties that they cannot be relied upon. Glanvill agreed that the fall had dimmed the faculty of reason, and therefore men should not be contentious about their opinions. If "reason" refers to the truths of reason, however, the eternal truths of nature, they are neither dimmed nor corrupted. They endure forever and are forever in harmony with religion. It is true that we may perceive them incorrectly and that men may dispute what they are, but men also misinterpret the Bible and dispute its true meaning. Human reason is not infallible, yet only a madman puts out his eyes because they are dim. "Reason," and here the influence of the Cambridge Platonists becomes manifest, "is, in a sense, the Word of God, *viz.* that which He has written upon our minds and hearts, as Scripture is that which is written in a book. The former is the word whereby He has spoken to all mankind; the latter is that whereby He has declared His will to the Church and His peculiar people."

Belief in reason, therefore, is an exercise of faith, and faith is an act of reason. Relying on reason in things clearly perceived is trusting in God's veracity and goodness, and that is an exercise of faith. "No principle of reason contradicts any article of faith. This follows upon the whole. Faith befriends reason, and reason serves religion, and therefore they cannot clash. They are both certain, both the truths of God; and one truth does not interfere with another. . . . Whatsoever contradicts faith is opposite to reason, for 'tis a fundamental principle of that that God is to be believed." At times the two do seem to contradict each other; something is taken for faith that is fancy, or something for reason that is sophistry. If something that seems to contradict reason is offered as an article of faith, we must investigate to assure ourselves that it is truly revealed by God. If it is revealed, the contradiction is only apparent, for God cannot be the author of contradictions. We ought not to accept anything as an article of faith, Glanvill concluded, "that palpably contradicts reason, no more than we may receive any in a sense that contradicts other Scriptures." [21]

> The denial of reason in religion has been the principal engine that heretics and enthusiasts have used against the faith and that which lays us open to infinite follies and impostures. . . . The impostures of men's fancies must not be seen in too much light, and we cannot dream with our eyes open. Reason would discover the nakedness of sacred whimsies and the vanity of mysterious nonsense; this would disparage the darlings of the brain and cool the pleasant heats of kindled

21. *Seasonable Recommendation . . . of Reason,* pp. 24–6.

> imagination. And therefore reason must be decried because an enemy to madness, and fancy set up under the notion of faith and inspiration. . . . Men have been taught to put out their eyes that they might see and to hoodwink themselves that they might avoid precipices. Thus have all extravagancies been brought into religion beyond the imaginations of a fever and the conceits of midnight. Whatever is fancied is certain, and whatever is vehement is sacred; everything must be believed that is dreamed, and everything that is absurd is a mystery. . . . Thus has religion by the disparagement of reason been made a medley of fantastic trash, spiritualized into a heap of vapors, and formed into a castle of clouds, and exposed to every wind of humor and imagination.[22]

The true intent of Glanvill's analysis of the scope of reason in religion is seen in his essay, "Anti-fanatical Religion and Free Philosophy," the account of the Utopian country of Bensalem. The Ataxites (sectarians and enthusiasts) of Bensalem made a great fuss over mysteries, built schemes of divinity out of absurdities and unintelligible fancies, and then counted their belief in such freaks as faith and spirituality. The new divines (Cambridge Platonists or Latitudinarians) replied with an explanation of the true sense in which the Gospel is a mystery—a secret that was hidden in the counsels of God and not discoverable by human inquiry until He was pleased to reveal it through Christ. Religion may yet be called a mystery because it is an art difficult to practice. Although the necessary articles of religion are revealed so clearly that they may be known by every sincere inquirer and have no obscurity in them in that respect, some of the propositions may be styled mysterious, because they are "inconceivable as to the manner of them." Thus the fact of the immaculate conception is perfectly clear, but its means are inexplicable. The new divines of Bensalem, with whom Glanvill classified himself, held that our faith is not concerned with the manner in which the things above reason are done; that belief in the simple article by itself is sufficient;

22. Ibid., pp. 30–2.

> that we are not to puzzle ourselves with contradictions and knots of subtlety and fancy and then call them by the name of mysteries; that to affect these is dangerous vanity, and to believe them is silliness and credulity; that by and on the occasion of such pretended mysteries the simplicity of the Gospel has been destroyed, the minds of men infatuated, sober Christians despised, the peace of the Church disturbed, the honor of religion exposed, the practice of holiness and virtue neglected, and the world disposed to infidelity and atheism itself.[23]

While he admitted that there were religious mysteries above human comprehension, he did not consider them to be of any importance. Disputes over them had torn society for too long while the practice of the few plain essentials had been ignored. The time had come to forget insoluble minutiae and to concentrate on the fundamentals which are clear to all men.

Joseph Glanvill claimed for the authority of human reason in religion a boundary beyond that marked out by Boyle. On the surface the positions of the two men were similar. Both admitted considerable scope to human reason, and both maintained that there are doctrines beyond human understanding. However, where Boyle contended that man is incompetent to judge spiritual truths delivered by revelation, that man should doubt his own reason before he denies revealed doctrines, Glanvill insisted that man should accept nothing inconsistent with what his reason concludes. Glanvill's greater emphasis on reason arose in part from his purpose. While Boyle was defending religious faith from the assertion that men cannot accept anything above reason, Glanvill was attacking enthusiasm, which embraced religious mysteries too eagerly. When he insisted that the articles of revelation cannot contradict reason, Glanvill meant the enthusiast's supposed personal revelation rather than the doctrines revealed in the Bible. Boyle and Glanvill would probably have found themselves in close agreement on the articles above reason that they accepted. Nevertheless a fundamental difference separated them. Boyle declared firmly that human reason is not the standard of

23. "Anti-fanatical Religion and Free Philosophy," *Essays*, Essay 7, pp. 29–30.

spiritual truth; Glanvill implied that it is. True, he admitted articles of faith above reason, but he would not agree to any that are contrary to reason. Boyle suspected the competence of the human intellect in case of conflict; Glanvill suspected the article of faith. Moreover, the very doctrines above reason that Glanvill agreed to accept were thrust into a corner and labeled unimportant. The significant part of religion is the part demonstrable by human reason, the simple fundamentals and the moral law. The tendency to emphasize the reasonable fundamentals at the expense of the mysteries of faith was clearly present in Boyle; Glanvill carried it to a more advanced stage. According to his analysis little of importance in religion stands beyond the range of human comprehension.

What Glanvill advanced so far, John Locke completed. In his analysis of faith and reason religion was brought almost entirely under the dominion of human understanding. Locke's whole instinct was set against the attitude of faith; such an imaginative delight in the enigmatic and the mysterious as characterized Sir Thomas Browne was alien and antithetical, even distasteful, to his mind. He was the rationalist, the searcher, the disciple of truth. Without approaching iconoclasm, Locke demanded of himself that he try all things and assent to nothing lacking substantial proof; and he thought that every man, to be worthy of the gift of reason, must follow the same course. To be sure, Locke did not believe that human reason can probe all things. The *Essay Concerning Human Understanding* (1690) attempted to decide in what matters reason can attain to knowledge. Perhaps the real Locke is seen more clearly in the *Conduct of the Understanding* (1706, in the *Posthumous Works*) urging men to pursue truth instead of pointing out the limits of certainty as he did in *Human Understanding*. "We are born with faculties and powers capable of almost anything," he proclaimed, "such at least as would carry us farther than can easily be imagined." [24] His purpose in analyzing the boundaries of certainty was not to encourage submission to faith but to direct men away from futile inquiries into the unknowable. Human efforts might then be turned to studies in which they would reap some useful harvest. If Locke did not

24. Locke, "Conduct of the Understanding," in *Works*, 2, 331.

contend that human understanding can comprehend everything, neither did he think that its compass is narrowly restricted. In reality he believed that human understanding can know everything that is worth knowing. Man's faculty of reason, Locke maintained, is the gift of God, the Almighty's noblest creation; to impugn or to ignore it is to disparage His wisdom. In the words of the splendid tribute to Locke written by Lady Masham, "He was always, in the greatest and in the smallest affairs of human life as well as in speculative opinions, disposed to follow reason, whosoever it were that suggested it; he being ever a faithful servant, I had almost said a slave, to truth, never abandoning her for anything else and following her for her own sake purely."[25]

Locke's purpose in examining the competence of human reason in religion, like that of Glanvill, was not to protect the area of faith, which he considered too large and well established among the majority of men already. Like Glanvill again, the spectacle of enthusiasm, revolting to his nature, appears to have been a major influence driving him into religious rationalism. When he wished, Locke could be bitingly scornful; for enthusiasm he had only scorn of the most biting kind. If such groundless delusions as the "illuminations" of enthusiasm possess a man's mind about ordinary matters, he declared, we call it raving and consider it a degree of madness; but men are accustomed to think they may and ought to abandon reason in religion.[26] To the crying up of faith at the expense of reason we must ascribe those absurdities that fill all of the religions which rule and divide mankind, he wrote in the *Essay Concerning Human Understanding*.

> For men having been principled with an opinion that they must not consult reason in the things of religion, however apparently contradictory to common sense and the very principles of their knowledge, have let loose their fancies and natural superstition, and have been by them led into so strange opinions and extravagant practices in religion that a considerate man cannot but stand amazed at their follies and judge them so far from being acceptable to the great and wise God, that he cannot avoid thinking them ridiculous

25. Cited in Fox Bourne, *Locke, 1,* opposite title page.
26. Bodleian, Locke MS f 6; journal entry for February 19, 1681/2.

> and offensive to sober men. So that in effect religion, which should most distinguish us from beasts and ought most peculiarly to elevate us, as rational creatures, above brutes, is that wherein men most often appear most irrational and more senseless than beasts themselves.[27]

Locke also remembered the great Civil War which had raged through the England of his youth. If men could only be brought to use their reason, to realize their agreement on the reasonable fundamentals, and to recognize the unimportance and uncertainty of the details separating them, society could be preserved from upheaval and religion saved from self-destruction.

In examining the relation of reason to faith Locke divided propositions into three categories distinguished by the degree to which they conformed to reason. Any proposition must be according to reason, above reason, or contrary to reason. According to reason are those propositions the truth of which can be traced from the simple ideas we gain through sensation and reflection. Above reason are those propositions which cannot be derived from our simple ideas. Contrary to reason are those propositions which are inconsistent with or irreconcilable to our clear and distinct ideas. Faith properly belongs to propositions above reason that are revealed by God. Locke made human reason the final arbiter even of those doctrines above reason, for he maintained that only by the exercise of reason can man decide whether they are truly revealed by God. Merely a strong conviction of the mind, an inner light, does not suffice to establish a revelation; no one can doubt from the many strange and contradictory opinions in the world that men are likely to have more fancies than inspirations. Miracles alone can prove that God has spoken. Miracles in turn are difficult to recognize, since we do not know the final limits of natural causes. In effect Locke was limiting true revelation to the Bible. He did not believe in latter-day miracles, and nothing in his writings encouraged reliance on personal revelation. His sense of tradition and his reluctance to employ his analysis of knowledge simply for criticism and destruction did not allow him to challenge the Bible, even when the scrupu-

27. *Human Understanding*, Bk. IV, chap. xix, sec. 11.

lous application of his own principles would surely have raised doubts about the Scripture's authority. To be sure, Locke was not unwilling to apply a limited test of reason to the Scriptures. In their interpretation he insisted that common sense must be the guide; they must be examined as rational and coherent discourses instead of tortured to find obscure truths in isolated texts. He even confessed that he could not believe that everything in the Bible has equal authority and inspiration.[28] To doubt the authority of the Bible as a whole, however, was a step which he would not take. Considering the Bible as the word of God confirmed by miracles, he was willing to accept on faith those truths above reason which it revealed.

Locke's definition of faith suggested, however, that it was inferior to real knowledge. Faith, according to him, "is a settled and sure principle of assent and assurance and leaves no manner of room for doubt or hesitation." [29] Knowledge gives certainty; faith gives assurance—Locke insisted on the distinction. "Faith stands by itself and upon grounds of its own, nor can be removed from them and placed on those of knowledge," he replied to the protesting Bishop of Worcester. "Their grounds are so far from being the same or having anything common that when it is brought to certainty faith is destroyed; it is knowledge then and faith no longer. With what assurance soever of believing I assent to any article of faith, so that I steadfastly venture my all upon it, it is still but believing. Bring it to certainty, and it ceases to be faith." [30] The implication of his analysis, distinguishing the assurance of faith from the certainty of knowledge, placed faith on a lower plane than the knowledge gained by rational inquiry.

Although Locke's implied derogation of faith made little practical difference in the acceptance of things above reason, it became decisively important when applied to his third category of propositions, things contrary to reason. In his opinion faith applies properly only to things above reason, while those who maintained that faith is a higher principle than reason held that revealed doctrines contrary to reason must also be accepted. Locke

28. Locke to Limborch, October 6, 1685; cited in Fox Bourne, *Locke*, 2, 32.

29. Locke, *Human Understanding*, Bk. IV, chap. xvi, sec. 14.

30. Locke, "Second Reply to the Bishop of Worcester," in *Works*, 3, 146–7.

refused absolutely to accept Boyle's contention that the human intellect cannot reject articles of faith merely because they seem to contradict reason. In all those things which have clear evidence from sound principles of knowledge reason is the proper judge, he declared; revelation may confirm reason, but it cannot overthrow its conclusions. Where we have the clear sentence of reason, we cannot ignore it on the pretense of faith, which has no authority against the plain and clear dictates of reason. No evidence that any revelation originates from God can be so certain as the principles of reason, and therefore nothing contrary to reason has the right to be urged as a matter of faith. For God to overturn the foundations of reason itself in order to set up a doctrine contrary to reason would be a greater miracle than any confirming a revelation. This would set one miracle against another with the greater miracle on the side of reason; the seeming miracle confirming the revelation in such a case would probably be a natural phenomenon the cause of which we do not know. We cannot possibly believe a doctrine to come from God, the bountiful Author of our being, Locke asserted, "which if received for true, must overturn all the principles and foundations of knowledge He has given us, render all our faculties useless, wholly destroy the most excellent part of His workmanship, our understandings, and put a man in a condition wherein he will have less light, less conduct than the beast that perishes." [31] Reason is natural revelation by which God communicates to mankind that portion of knowledge which He has laid within the reach of their natural faculties. Revelation is natural reason enlarged by a new set of discoveries, communicated immediately by God, which reason confirms by the testimonies and proofs that it comes from God. "So that he that takes away reason to make way for revelation puts out the light of both, and does much what the same as if he would persuade a man to put out his eyes, the better to receive the remote light of an invisible star by a telescope." [32]

"Light, true light in the mind," said Locke, "is or can be nothing else but the evidence of the truth of any proposition; and if it be not a self-evident proposition, all the light it has, or can

31. Locke, *Human Understanding,* Bk. IV, chap. xviii, sec. 5.
32. Ibid., Bk. IV, chap. xix, sec. 4.

have, is from the clearness and validity of those proofs upon which it is received."[33] The full impact of Locke's analysis of reason and faith depends upon his conception of reason which denied the reality of "light in the mind" in the sense of truths printed directly upon the human intellect. His discussion of reason and religion, appearing in the fourth book of the *Essay Concerning Human Understanding,* applied the principles worked out in the first three books. Categorically denying the existence of innate ideas, Locke held that all of our ideas are based on sense perception and reflection on our own mental processes. We can have knowledge only so far as we have such ideas. Pure rationalism was foreign to Locke's mind. He insisted that knowledge must return continually to its empirical foundation, and that human reason cannot pretend to consider matters totally removed from the simple ideas presented to it by the senses. For too long philosophy had been a game played with empty words. Men tried to encompass the whole universe within their minds. They wasted their energy and time on useless and fruitless investigations of questions beyond human powers, while they ignored the things that human reason can know with certainty. To cover their ignorance they invented endless subtle distinctions without counterparts in reality and employed words that did not have clear ideas behind them. The litigious and disputatious philosophy of the Schoolmen had only confused men's minds without increasing their knowledge. Rejecting what he believed to be the errors of the past and insisting that human reason not rise too far from its empirical foundation, Locke used "reason" to mean the faculty by which men compare and combine the simple ideas derived from the senses. His contention that knowledge arises from sensory experience systematized the virtuosi's usual understanding of "reason," giving philosophical justification to their method of investigating nature. His conclusion on the role of reason in religion also completed the virtuosi's analysis of that subject, declaring that religion cannot include articles of belief that contradict the testimony of the senses.

In addition to meaning the faculty that manipulates simple ideas, "reason" also meant to Locke the body of conclusions legit-

33. Ibid., sec. 13.

imately drawn from the simple ideas. Used in this sense, "reason" involved a definite conception of the external world depicted by sensory perceptions. To Locke external nature was the cosmic machine commonly accepted by the virtuosi. "Reason," in this meaning, was another facet of the protest against the Scholastics which pervaded both Locke's works and those of the virtuosi at large. The Scholastics had built castles in the air; instead of describing nature, they had dealt in verbal distinctions which were not related to actual existing things. Boyle spoke for every virtuoso when he condemned "those curious but groundless structures that men have built up of opinions alone . . ." [34] Scholastic philosophy was misdirected as well, pursuing the wrong questions with the wrong notions. Glanvill's doggerel couplet in *Plus Ultra* epitomized the virtuoso's picture of the Scholastic philosopher:

> He knows what's what, and that's as high
> As metaphysic wit can fly.[35]

In opposition to Scholasticism the virtuosi raised the demand for "real and useful knowledge"—real in explaining the actual, that is, the mechanical, world of nature, useful in ministering to material needs. The phrase appeared repeatedly in their writings. Locke fully agreed with the virtuosi's condemnation of Scholasticism, so that "reason" in his mind involved the virtuosi's conception of nature. Reason was descriptive rather than explanatory; it concerned itself with concrete facts perceptible to the senses while it turned away from the investigation of ultimate meanings. Thus Locke's rejection of things contrary to reason carried a special implication. It meant in part that no article of religion could contradict the conclusions of natural philosophy. It assumed that all of God's acts conformed to the pattern of mechanical nature.

Locke borrowed from the Scholastics the distinctions of things according to reason, things above reason, and things contrary to reason. With them it had been a justification of things above reason. Locke took the words and informed them with the spirit of empirical rationalism, converting the formula into an attack on

34. Boyle, *Works, 1,* 312.

35. Joseph Glanvill, *Plus Ultra: or, the Progress and Advancement of Knowledge since the Days of Aristotle* (London, 1668), p. 120.

enthusiasm and on all of the seemingly obscure Christian doctrines that the centuries had produced. Although he admitted the category of things above reason, he redefined its meaning and left nothing in it corresponding to the former meaning. When Locke discussed things above reason in the *Essay Concerning Human Understanding,* he offered two examples. That part of the angels rebelled and thereby lost their first happy state, and that the dead shall rise and live again, propositions beyond the discovery of human reason, are purely matters of faith with which reason has nothing to do directly.[36] Another example might be what the *Reasonableness of Christianity* considered to be the essential doctrine of Christianity, that Jesus was the promised Messiah. These are doctrines above reason in the sense that human reason alone cannot discover them; no sense data pertain to them. They are not, however, spiritual mysteries incomprehensible to man. Such doctrines as the Trinity and the hypostatical union of God and man in Christ belong to a different order, things that man cannot understand even when revealed, things contradictory to physical experience. Locke rejected the Trinity and the divinity of Christ. Since he excluded the articles of faith that traditional Christianity had considered to be above reason, his category of doctrines above reason was little more than a gesture simulating orthodoxy. It did not lift his conception of Christianity beyond the judgment of the unaided faculties of man.

Locke's discussion of reason and faith should be read in the light of his *Letters on Toleration.* The care of every man's soul belongs to himself alone, he asserted; no authority can be allowed to invade the individual's right of private judgment in a matter of the utmost personal concern. To contend that religion is entirely a personal affair was to undercut the ground of faith still further. Whatever the proper boundary between faith and reason, he was saying, no authority can establish it. There can be no authoritative body of doctrines to be accepted on faith. The Bible is the source of Christian truth to be sure, but in the end each individual must interpret the Bible for himself. Thus in maintaining the rights of human reason in religion Locke was maintaining the right of each individual to judge the articles of faith by his

36. Locke, *Human Understanding,* Bk. IV, chap. xviii, sec. 7.

own reason. He was denying any practical means to determine the articles of faith conclusively.

In his Journal for 1677 Locke composed a long essay on the limits and ends of human knowledge which casts a good deal of light both on the purpose of the *Essay Concerning Human Understanding* and on his religious opinions. The inner structure of things and the vast frame of the universe, Locke began, are quite beyond human comprehension; efforts to penetrate these areas are as futile as they are useless. If we consider our condition in the world, we find that we are in need of food, clothing, and shelter, which must be supplied by our labor; by art and knowledge we are able to supply these things more easily. Man is capable of such knowledge, and he should apply his capacities to the improvement of useful arts. Man is also destined to a future life if he guides his actions aright, and his abilities are sufficient to distinguish the proper course. Our minds, then, are able to lead to happiness here and in the hereafter; we should use them for the purpose for which they are competent; we should stop wasting our time on speculations about the nature of the universe, which can never reach any conclusion and which would be of no value if they did.[37] "Our business here," he declared in the opening passages of the *Essay Concerning Human Understanding,* "is not to know all things, but those which concern our conduct. If we can find out those measures whereby a rational creature, put in that state in which man is in this world, may and ought to govern his opinions and actions depending thereon, we need not be troubled that some other things escape our knowledge."[38] As he was impatient with philosophical examination of such concepts as substance, so in religion he had no stomach for esoteric doctrines. He thought that human reason can discover the fundamental and necessary doctrines; to seek to build theological systems above the foundations is fruitless. All other points except the fundamentals are conjectural in the extreme; and not only are they conjectural, they are unimportant as well. The existence of God can be demonstrated, and from the existence of God the moral law can be deduced. Locke contended that morality—that

37. Bodleian, Locke MS f 2; journal entry for February 8, 1677.

38. Locke, *Human Understanding,* Bk. I, chap. i, sec. 6.

is, obedience to divine law—is the true essence of religion. Theological systems are wonderfully productive of disputes, even of wars and persecution; but in the smoke of battle the moral law is more often than not lost from sight. Locke sought to bring religion back to its reasonable fundamentals, back to the things that count, away from useless speculations about insignificant theological distinctions. In the *Reasonableness of Christianity* he urged that only the acceptance of Christ as the Messiah is essential to Christianity.

> This is a plain intelligible proposition; and the all-merciful God seems herein to have consulted the poor of this world and the bulk of mankind. These are articles that the laboring and illiterate man may comprehend. This is a religion suited to vulgar capacities, and the state of mankind in this world, destined to labor and travail. The writers and wranglers in religion fill it up with niceties and dress it up with notions, which they make necessary and fundamental parts of it, as if there were no way into the church but through the academy or lyceum.[39]

Locke utterly rejected speculation about obscure points; religion as he conceived of it fell entirely within the powers of human understanding.

The pattern of development of the four virtuosi's thought on the place of reason in religion necessarily followed the course of natural religion's growing importance among them. The different purposes of their analyses reflected their different approaches to natural religion. Both Wilkins and Boyle wanted to assert the limits of man's ability to comprehend religious truths even while they affirmed reason's validity within those limits. Although they examined somewhat different problems, both agreed in the conclusion that religion cannot be judged entirely by the standards of philosophy since religion is concerned with the spiritual realm where material data do not apply. This was an aspect of the attack on atheism and skepticism for which they employed natural religion, and the limitation of human reason's competence that they suggested reflected their belief that natural

39. Locke, *Works*, 6, 157.

religion was only the foundation of Christianity. As natural religion, to which the virtuosi initially resorted as a rational defense of Christianity, culminated in virtually displacing Christianity in their religious thought, so the analyses of reason's competence changed from negative indication of limits to positive assertion of authority. Glanvill and Locke did not look upon the faculty of reason as an instrument of limited value, fit only to confirm Christianity but not to question its doctrines. All religion must be examined by reason to expose and destroy absurdities and to save religion from its own excesses. With Locke, Christianity became practically identical with natural religion, and with Locke reason was granted full scope to consider and to judge nearly all religious matters. Natural religion became the religion of reason, Christianity pruned of irrational doctrines. None of the four used reason as a tool of systematic skepticism. If some of them questioned and rejected specific doctrines, they did not question religion itself. Their function, as they thought of it, was to purify and reform. Even in rejecting traditional doctrines of orthodoxy, Locke thought that he was serving the final interest of Christianity. The virtuosi's belief that natural philosophy could not harm Christianity was reflected in their belief that the exercise of reason could not invalidate religion.

Fundamentally the examination of reason's scope in religion was an examination of human capacity. Although they differed in their definitions of reason, the virtuosi agreed that God and the creation are rational; the question to be decided was how far human reason can penetrate universal reason. The increasing emphasis on reason in religion represented a growing confidence in human powers. The virtuosi became increasingly reluctant to accept doctrines contrary to or even above the conclusions of human reason. They tended more and more to treat doctrines above reason as inconsequential minutiae fit only to disturb the leisure of cloistered monks. In the fullness of confidence they were ready to deny that anything important transcended their capacity. Neither disproving nor rationalizing mysteries, they simply rejected them, thus making an essential contribution to the growth of natural religion.

CHAPTER 8

Isaac Newton: A Summation

> The main business of natural philosophy is to argue from phenomena without feigning hypotheses, and to deduce causes from effects, till we come to the very first cause which certainly is not mechanical . . .
>
> Newton, *Optics*

> It is the temper of the hot and superstitious part of mankind in matters of religion ever to be fond of mysteries, and for that reason to like best what they understand least.
>
> Newton, "An Historical Account of Two Notable Corruptions of Scripture"

"WHEN I WROTE my treatise about our system," Isaac Newton began a letter to Richard Bentley, "I had an eye upon such principles as might work with considering men for the belief of a deity; and nothing can rejoice me more than to find it useful for that purpose."[1] If the direct reference to the *Principia* be excepted, Newton's opening sentence in reply to Bentley's request for aid in preparing the Boyle Lectures might well have been the words of any virtuoso of the late 17th century. As Newton the scientist summed up and in a sense completed the 17th century's progress in scientific knowledge, so also Newton the Christian virtuoso may be seen as a synthesizing figure. Standing at the end of the century, he summarized the religious attitudes that had developed among the virtuosi as they had hammered out their reconciliation of Christianity and natural science. His position in the religious thought of the virtuosi is different, of course, from his position in their scientific thought. In science Newton made momentous new discoveries which have placed his name

1. Newton to Bentley, December 10, 1692; Isaac Newton, *Four Letters from Sir Isaac Newton to Doctor Bentley. Containing Some Arguments in Proof of a Deity* (London, 1756), p. 1.

foremost among the world's scientists; in religion, although he labored over his manuscripts with painful care, he produced nothing exciting and little new. Where in natural science he was a beacon guiding the way, in religion he was only a mirror reflecting conclusions reached by others. As a mirror, however, Newton did faithfully reproduce the religious attitudes of the Christian virtuosi, so that he presented a summary and a culmination of their religious thought, the product of the interaction of science and religion within the framework of influences that characterized the late 17th century.

The relationship within Newton's own mind between his scientific work and his religious beliefs was a complex network of mutual influence. The traditional Christian persuasions, which Newton was reared to accept, were neither wholly displaced nor wholly untouched, while his scientific theories were not unaffected by Christian doctrines. In fine, Newton, like the other virtuosi, effected a compromise between natural philosophy and the traditional Christian view of the world and of life. Although he went a step beyond the others in forcing Christianity into conformity with science, his compromise was still essentially the one accepted by all the virtuosi.

The harmony of science and religion was the basic proposition. When Newton wrote to Bentley that his *Principia* had kept an eye to principles pointing to the existence of God, he was expressing the standard conviction of the virtuosi. As seen through their eyes, natural philosophy investigated the same ultimate truth which Christianity revealed. Somewhat like the medieval Scholastics Newton placed science at the service of religion. Unmoved as he was by the Baconian spirit which animated so many of his fellows, the urge to use science as a tool to dominate and to control nature, Newton stated that the primary goal of scientific investigation is to reveal the ultimate cause of creation. The main business of natural philosophy, he declared in the *Optics,* is not to unfold the mechanism of the world, but to deduce causes from effects until we arrive at the First Cause, which is certainly not mechanical. While every step does not immediately reveal the First Cause, it brings us closer to knowledge of it, "and on this account is to be highly valued."[2] Although the contention that the creation reveals the Creator had become almost trite, a genuine feeling of religious awe before the wonders of creation lay behind the bare idea. With the other virtuosi Newton shared the feeling to the fullest degree. "O unprofuse magnificence divine!" the poet James Thomson exclaimed in response to the universe revealed in Newton's *Principia,*

> O unprofuse magnificence divine!
> O wisdom truly perfect! thus to call
> From a few causes such a scheme of things,
> Effects so various, beautiful, and great,
> An universe complete![3]

The spirit of Thomson's lines aptly portrayed Newton's own response, so that it required no wrenching of his principles to declare that he valued science chiefly as it brought him nearer to knowledge of the First Cause. With the other virtuosi he believed that science was naturally in harmony with religion.

Every virtuoso found particular evidence of divine handicraft in his own special field of study. John Ray discovered the imprint of God in all forms of life, Richard Lower in the design of the heart, Nehemiah Grew in the structure of plants. Newton approached God most directly in contemplating the system of the universe. When Richard Bentley was preparing the Boyle lectures in defense of religion in 1692, he wrote to Newton for advice. In the correspondence that ensued, Newton explained how nature revealed her Creator to him. Let us assume, Newton began, that matter was originally diffused evenly through space. If space is finite, then gravity would have brought all matter together into one body; if infinite, into a number of bodies. The universe that exists suggests the latter supposition. If we assume further that matter is luminous, natural causes alone might explain the formation of the sun and the stars. Natural causes alone cannot, how-

2. Isaac Newton, *Optics: or, a Treatise of the Reflections, Refractions, Inflections and Colors of Light* (3d ed. London, 1721), p. 345.

3. James Thomson, *A Poem Sacred to the Memory of Sir Isaac Newton* (London, 1727), pp. 8–9.

ever, explain how luminous and nonluminous matter divided themselves into separate bodies; to arrive at the system of the universe in its full complexity we are forced to resort to an intelligent agent. The same agent placed the sun in the center of the planets; for if only blind chance were at work, the sun ought to be, like the planets, dead and devoid of light and heat. Why should there be a body in the system qualified to give light and heat to the rest? Solely because God thought it convenient, Newton replied. Why only one? Chance might have provided many, but intelligence saw that one was sufficient. Intelligent design was also necessary to make the planets move in the same direction, in the same plane, and at such speeds that their orbits are nearly concentric. The work of chance is displayed in the unplanned and eccentric orbits of comets. When contrasted to the intelligent order of the planetary system, comets demonstrate that natural causes alone could not have produced the universe.

> To make this system, therefore, with all its motions, required a cause which understood and compared together the quantities of matter in the several bodies of the sun and planets and the gravitating powers resulting from thence, the several distances of the primary planets from the sun and of the secondary ones from Saturn, Jupiter, and the earth, and the velocities with which these planets could revolve about those quantities of matter in the central bodies; and to compare and adjust all these things together in so great a variety of bodies argues that cause to be not blind and fortuitous, but very well skilled in mechanics and geometry.[4]

Very well skilled in mechanics and geometry—the master of mechanics and geometry was paying his respects to the One Who excelled him. Newton, who could brook no human critic, acknowledged the work of his superior. All of his own skill had been necessary to plumb the system of the universe. How much greater must be the intelligence of Him Who created it! Christian virtuoso that he was, Newton stood in reverent wonder before the cosmic machine.

On the surface Newton's correspondence with Bentley ex-

4. Newton to Bentley, December 10, 1692; Newton, *Letters to Bentley*, pp. 7–8.

pressed the spontaneous reverence of a virtuoso contemplating nature; in its assumptions it revealed that Newton was admiring an order projected upon nature by his own mind. He was reading purpose and design into a set of empirical facts. The simplicity of the solar system, with all of the planets circling the sun in the same plane, seemed to Newton a token of divine intelligence; but this was to assume that simplicity is a divine end. Might God not manifest Himself equally well in utilizing a different plane for each planet, making all of the planes serve one solar system? John Ray, for example, found a sign of divine omnipotence in the multiplicity of creatures. Newton imagined design in the existence of a single sun warming the planets; had the numbers been reversed with seven suns warming one planet, he would undoubtedly have perceived the hand of God in that arrangement as well. The fact is that Newton was convinced from the beginning that the universe is an ordered cosmos because he knew as a Christian that God had created it. An exchange of letters in 1698 with an Oxford undergraduate, John Harington, affords another glimpse of Newton's fundamental assumptions. Harington found a sympathetic audience when he wrote to Newton of his theory that harmonic ratios are related to the proportions of good architecture. It seemed appropriate to Newton that God had formed the cosmos on the principles of geometry; therefore the geometric ratios ought to be found throughout nature. In his reply to Harington he encouraged the young man in the study which, as he said, "tends to exemplify the simplicity in all the works of the Creator." [5] Here the prior conviction of order and simplicity was dictating a line of investigation. The letter to Harington explains Newton's fascination with a pet theory that the ratios between the spaces occupied by the colors in the spectrum reproduce the harmonic ratios of the octave. Despite the fact that the correspondence between the spectrum and the octave is not exact, Newton toyed with the idea for years, apparently convinced that he was grasping at the fundamental structure of the cosmos. Like the other virtuosi

5. Newton to Harington, May 30, 1698; Isaac Newton, *Thirteen Letters from Sir Isaac Newton, Representative in Parliament of the University of Cambridge, to John Covel, D.D., Vice-Chancellor, Master of Christ's College* (Norwich, 1848), p. 303.

Newton came to the investigation of nature with presuppositions drawn from Christianity which colored his idea of nature. The study of natural philosophy neither strengthened nor weakened his belief in God, but it offered a medium through which his piety could express itself.

The piety of Newton before the works of nature summarized the traditional element which remained undisturbed in the religion of the virtuosi. To them the world was more than an autonomous machine thrown together by chance. The well-wrought creation of God, nature appeared to them as a display of unparalleled creative skill which humbled the proud investigator's mind. Not long before his death Newton reviewed his life for the husband of his favorite niece. "I do not know what I may appear to the world," the aged virtuoso mused; "but to myself I seem to have been only like a boy playing on the seashore and diverting myself in now and then finding a smoother pebble or a prettier shell than the ordinary, whilst the great ocean of truth lay all undiscovered before me." [6] Robert Boyle used the same figure, comparing the virtuoso to a man drifting down a river; the further he goes, the more the shores recede until upon reaching the river's mouth he finds himself before a boundless sea. The Christian virtuoso of the 17th century was that boy on the shore, that man at the river's mouth, spellbound and struck with wonder before the limitless ocean of truth.

In the history of religion, however, the role of the virtuosi extended beyond the preservation of piety. Not all of Christianity survived their touch so well. Despite their worshipful frame of mind, there were aspects of their work which challenged received doctrines of religion. Indeed their very inquiries into nature helped to render the natural order which inspired them with such wonder considerably less wonderful to those who followed them. Before Newton lived, comets were exciting rarities; he proved that they were ordinary components of the universe, bound by the laws that govern all bodies. His law of gravitation reduced the complexity of the heavens to a relatively simple and comprehensible order. A mystery probed is a mystery no longer, and Alexander

6. Portsmouth MSS.; cited in Louis Trenchard More, *Isaac Newton, a Biography* (New York, Scribner's, 1934), p. 664.

Pope said more than he intended, perhaps, when he exclaimed, "God said, 'Let Newton be,' and all was light." Medieval architects wisely stained the glass of their cathedral windows to tone down the direct rays of the sun. Newton himself believed that his system of the universe raised the fundamental question of its origin; a future generation might conclude that the mechanics of nature explain everything. Newton asked why the planets are so regularly arranged; a future generation might decide that gravity is reason enough. Newton looked with reverence on the creation, the laws of which he had labored to discover; a future generation which received the system intact without the labor might take it for granted. There was a contradiction between the virtuosi's attitude of wonder and their efforts to inquire and explain, and in their investigations they helped to eliminate the element of mystery which they worshiped.

The impact of science upon Christianity may also be traced in the virtuosi's picture of God. While they asserted the harmony of science and religion, they were bending their conception of God to fit into their scientific universe, until with Newton the Father was fading into a metaphysical projection of the creation. The nature of God was adjusted to the nature of the universe as science revealed it. In the General Scholium of the *Principia*, for example, Newton nearly identified the Almighty with the eternity and infinity of the world. "He endures forever and is everywhere present," he declared; "and by existing always and everywhere, He constitutes duration and space." [7] In one of his theological papers he even began to assign physical properties to God; the Father is immovable, he said; the Father is invisible.[8] The extent to which scientific categories dominated Newton's picture of God emerges suddenly from an unguarded comment in a letter. Writing to Dr. Thomas Burnet about the Mosaic account of creation, Newton tried to explain the seven days of creation, a problem which bothered everyone who wrote about the creation. Could God have made everything in six ordinary days? If not, what do the "days" mentioned in Genesis signify? Newton's answer involved

7. Isaac Newton, *The Mathematical Principles of Natural Philosophy*, trans. Andrew Motte (2 vols. London, 1729), 2, 390.

8. Kings College, Keynes MSS, No. 8.

the simple application of mechanics. Before the earth attained its present speed of rotation, it had had to be accelerated from rest; the first days of creation were much longer because the earth was rotating more slowly. As the earth was gradually accelerated under steady force, it finally reached its present velocity, at which point the force was withdrawn, leaving it to rotate forever at the rate of three hundred and sixty-five revolutions per year.[9] Newton's explanation of the seven days of creation was plausible enough, but the interest of the explanation lies in the startling spectacle which it presents of God as a mechanical force, bound, even in the act of creation, to the universal laws of the *Principia*. Acceleration, Newton seemed to be thinking as he contemplated the earth's initial rotation, is proportional to (divine) force. The influence of scientific thought was having its effect; God was assuming the characteristics of His creation. Newton believed that the universe requires a Creator, but he did not see that a God deduced from nature can be no more than a projection of nature.

The impact of science on religion was not indirect and unconscious in every case. There was at least one area in which the virtuosi themselves saw the possibility of conflict and hence the necessity of accommodation. This was the apparent discrepancy between the mechanical conception of nature and the traditional Christian doctrine of providence. Every thoughtful virtuoso had to face the question; Newton was no exception. The problem as he confronted it at the end of the 17th century was identical with the problem that bothered earlier virtuosi, for Newton's extraordinary contributions to science did not alter the conception of nature that they had elaborated. The mechanical idea of nature had been fully worked out before his time; he accepted it without question and made it the basis of his work. The principal thing that he added to it was mathematical precision. Where earlier virtuosi, such as Robert Boyle, had explained the mechanical conception of nature with metaphors and analogies, Newton solidified the argument with exact mathematical calculations. There had been much talk about machines but little

9. Newton to Burnet, no date on letter, but written early in 1681; in Sir David Brewster, *Memoirs of the Life, Writings, and Discoveries of Sir Isaac Newton* (2 vols. Edinburgh, 1855), 2, 453.

knowledge of mechanics. Newton gave the basic principles of mechanics their classic formulation. After the publication of the *Principia* not only could a virtuoso argue that the universe operates like a machine, but he could calculate exactly the behavior of the parts when subject to given mechanical forces. In short, mathematical precision consolidated the ground already won from the Aristotelian foe by rational conjecture. Newton's scientific work made the universe, as it was held in theory, neither more nor less mechanically inexorable. Although Newton never formally undertook to reconcile his mechanical idea of nature with providence and left no fully elaborated statement, he did indicate his opinion in a number of brief passages.

In a famous jibe the German philosopher Leibniz charged that Newton pictured God as a bumbling watchmaker, so unskillful that His piece had to be cleaned and repaired from time to time.[10] Newton had suggested that small inequalities in planetary motion would increase until God would have to set the orbits back in order. In Leibniz' opinion this was to declare God to be incompetent; an omniscient Being should have been able to construct a machine that would not need repairs. Although Newton himself did not reply directly to Leibniz, Dr. Samuel Clarke took up the defense. Denying the whole clock analogy, he asserted that God did not leave the world to run by itself after He had created it; the Almighty continues actively to govern and to direct it at all times. Although Clarke's reply evidently had Newton's approval,[11] Leibniz' jibes came nearer to the spirit of Newton's writings. Newton's position on the dominion of God over the creation cannot be equated with the Christian doctrine of providence.

When he published the second edition of the *Principia* in 1713, Newton added a General Scholium, in which he answered Leibniz' criticisms without mentioning the German philosopher directly. God, Newton said, "governs all things, not as the soul of

10. Samuel Clarke, *A Collection of Papers Which Passed between the Late Learned Mr. Leibniz and Dr. Clarke in the Years 1715 and 1716. Relating to the Principles of Natural Philosophy and Religion* (London, 1717), p. 5.

11. Brewster states that Clarke received assistance from Newton on several astronomical points, and that papers among Newton's manuscripts contain the same views as those given by Clarke (Brewster, *Newton,* 2, 287). I have not seen those particular papers.

the world, but as Lord over all; and on account of His dominion He is wont to be called Lord God, Pantocrator, or Universal Ruler. For God is a relative word and has a respect to servants; and Deity is the dominion of God, not over His own body as those imagine who fancy God to be the soul of the world, but over servants. The supreme God is a Being eternal, infinite, absolutely perfect; but a being however perfect without dominion cannot be said to be Lord God . . . It is the dominion of a spiritual being which constitutes a God." [12]

On the surface Newton would seem to have had a full blown concept of divine providence, but in effect he did no more than substitute the word "dominion" for the word "providence"; and as Newton defined "dominion" he manifestly did not mean direct and immediate governance. God's dominion was His arbitrary freedom in shaping matter and promulgating laws at the original creation. Where Leibniz asserted that the act of creation was controlled by rational necessity, that God could have created only the most perfect of worlds, Newton replied that in exercising His dominion, God created laws of nature by an act of will unconstrained by rational necessity. Had He so willed, He could have created a different system. "Dominion" had a further meaning for Newton: it included the idea that God shaped raw matter into the frame of nature. That is to say, in upholding God's dominion Newton was rejecting the notion that matter in motion, governed by natural laws but without intelligent guidance, could have formed itself into an organized universe. In sum, then, Newton's use of the term "dominion" limited its application to the Almighty's absolute power over matter in the act of creation. Even the passage quoted from the General Scholium, which affirmed the dominion of God so forcibly, went on to define dominion through its contrast to necessity. A God, he declared, "without dominion, providence, and final causes is nothing else but fate and nature. Blind metaphysical necessity, which is certainly the same always and everywhere, could produce no variety of things. All that diversity of natural things which we find suited to different times and places could arise from nothing but the ideas and will of a Being necessarily existing." [13] Every struggle

12. *Principia,* 2, 389–90.
13. Ibid., p. 391.

to break the fetters of mechanical necessity only served to enmesh Newton more firmly in them. The more he tried to define providence, the more it became indistinguishable from the original act of creation in which the universal machine was constructed. In the beginning God created the laws of nature and formed matter into the visible universe. His continued dominion, if it can be called such, can only be His support of the laws that He promulgated. Some passages implied that Newton believed in the Creator's power to remake the world in a new form; such an act would not exceed the competence of His dominion. Until He chooses to do so, however, the universe operates without His direct intervention except on possible rare occasions. The decisive text is hidden in the thirty-first query of the *Optics*. After denying that the world might have arisen from chaos by the operation of natural laws alone, Newton added that "being once formed, it may continue by those laws for many ages."[14] Despite all his attempts to retain a meaningful idea of providence, Newton could not free himself from the position to which nearly all of the virtuosi were driven. Indeed there seems to have been no possible alternative. If the mechanical universe is a reality, as Newton firmly believed, providence can only mean God's concurrence in the operation of its laws. Newton's attempt to assign specific functions to God, such as correction of inequalities in the motion of planets, deserved Leibniz' taunt. Interplanetary plumbery of this sort bore no relation to providence as Christianity had taught it. The geared wheels of the cosmic machine left no place for a personal direct providence of the Almighty.

The few scattered passages in which Newton considered miracles throw some light on his attempts to deal with the broader issue of providence. In a purely religious context he did not question miracles. A letter to John Locke affirmed that "miracles of good credit continued in the Church for about two or three hundred years."[15] When he looked at miracles as a scientist, however, he began to draw distinctions and to make reservations. He treated them directly only in a short note which has survived among his papers. "For miracles are so called," it begins, "not be-

14. Pages 377–8.

15. Newton to Locke, February 16, 1691/2; cited in King, *Locke, 1,* 409. *Cf.,* Newton to Locke, May 3, 1692; cited in More, *Newton,* p. 370.

cause they are the works of God but because they happen seldom and for that reason create wonder. If they should happen constantly according to certain laws impressed upon the nature of things, they would be no longer wonders or miracles but would be considered in philosophy as part of the phenomena of nature notwithstanding that the cause of their causes might be unknown to us."[16]

The passage is ambiguous. It could mean that miracles are the immediate works of God's hand, while all natural phenomena, which are also His works, are produced by created agents.[17] It seems to mean, however, that what we call miracles are only rare events, the causes of which we do not know. The spirit of the latter interpretation dictated his letter to Dr. Thomas Burnet, author of the *Sacred Theory of the Earth.* While arguing that the Mosaic account of the creation tallies with the best scientific knowledge, Newton managed to explain away nearly every element of the miraculous in it. Even the deluge was ascribed to natural causes.[18] In short, Newton both believed in and did not believe in miracles. As a Christian he accepted them, and he did not openly attack them in any context. On the other hand, as a scientist he introduced qualifications which amounted to repudiation. He could not really tolerate the thought that God might upset the laws of nature which He had created. Although logic seemed to require the rejection of miracles, he could not bring himself finally to cast out a belief which he considered vital to Christianity; and he tried to save himself by hiding behind a façade of words. Repeating his ultimate conclusion on providence as a whole, he abandoned himself to ambiguities and inconsistencies, which gave the appearance of divine participation in nature, but not the substance.

In a study entitled *Matter and Gravity in Newton's Physical Philosophy* A. J. Snow argues that Newton had two hypotheses for the cause of gravity. In some discussions he seemed to say that gravity is caused by a material ether, but most of his writ-

16. Printed in More, *Newton,* p. 623.

17. Clarke took this position in his exchange with Leibniz: Clarke, *Collection of Papers between Leibniz and Clarke,* pp. 353–65.

18. Newton to Burnet, no date on letter, but written early in 1681; cited in Brewster, *Newton,* 2, 453.

ings, especially the General Scholium attached to the *Principia* and the thirty-first query at the end of the *Optics,* pointed to an immaterial cause. Newton felt, Snow continues, that if gravity were reduced to a physical explanation, God would be ruled out of the universe. This he would not accept. While Descartes conceived of God as the original cause of the mechanical order, Newton held that the regularity of nature is an actual and immediate product of God's activity. Thus the "ethereal spirit" by which Newton tried to account for the cause of gravity referred to an immaterial force, the direct activity of God Himself.[19] Despite the general excellence of Snow's work, his basic proposition cannot be accepted. Whenever Newton discussed the ethereal spirit, he used terms that apply only to a material ether. For instance, he assigned it material properties such as elasticity and parts. While it is true that Newton did not believe that his hypothesis was the definitive word on the cause of gravity, he did seek gravity's cause within the physical universe. Snow's thesis must be rejected; but even if his contention be granted and it be allowed that Newton did try to save the participation of God in the creation by assigning gravity to His direct causation, the difficulties besetting the doctrine of providence still remain acute. The Creator is still confined to upholding the general order of nature.

Although Newton's treatment of providence was less thorough and profound than Boyle's or Ray's, he illustrated the position at which half a century of discussion had arrived. First, the virtuosi could not reconcile the mechanical conception of the universe with the traditional Christian doctrine of particular providence. Second, however, they refused to let the doctrine of providence go. Unwilling to accept the idea that natural science might be incompatible with Christianity, they held on to the original terminology and sought a verbal formula in which words would take the place of substance. One after another they searched for the statement that would assimilate the opposites, affirming the mechanical idea of nature without explicitly denying God's active governance of creation. In the General Scholium, which

19. A. J. Snow, *Matter and Gravity in Newton's Physical Philosophy* (London, 1926), pp. 137–68.

dated from 1713, Newton did not advance beyond the position that Boyle had reached nearly fifty years earlier. Nevertheless, he did try. He could not let the question rest. The ultimate result of the virtuosi's consideration of providence, as seen in Newton's statements on the subject, amounted to an implicit renunciation of particular providence together with a strong reaffirmation of general providence. If the virtuosi's intellectual honesty allowed them no other conclusion, their sincerity as Christians prevented a more radical departure from orthodoxy. They asserted that the world is not an autonomous machine; they affirmed the predominance of spirit over matter in the creation. In the eyes of the virtuosi their formulation of the idea of providence was not a repudiation of traditional Christianity; it was a successful demonstration that Christian teachings were compatible with the new theories of natural philosophy.

Like other virtuosi Newton usually limited his discussion of providence to God's relation to the physical world. To some extent this treatment of providence arose from the virtuosi's scientific work in which the mechanical hypothesis questioned the possibility of divine interference with the laws governing matter. On the other hand the virtuosi's efforts to arrive at a satisfactory statement of God's relation to the material world reflected the emphasis of their whole religious thought. The virtuosi were concerned more with intellectual systems than with emotional experiences; their religion aimed at the head instead of the heart. In this respect once again Newton summarized their position. In his search for a rational formulation of Christianity that would save both his religion and his faith in reason, he ignored the spiritual teachings that had in the past ministered to the soul of man.

The virtuosi originally turned to natural religion to demonstrate the rational foundation of Christianity, and most of their religious writings were devoted to the proof of the fundamentals of religion from the world of nature. With Newton the growth of natural religion reached its culmination. Early virtuosi thought of it as the foundation demonstrable by reason on which the superstructure of revealed Christianity rests. Newton followed Locke in confounding the superstructure with the foundation and embracing natural religion as the whole of Christianity.

The clearest expression of Newton's natural religion is found in a manuscript treatise entitled "Irenicum." That he considered the treatise important is attested by the labor he devoted to it. As he continually revised and reworked it, he gave more attention to the "Irenicum" than to any other theological writing: nearly ten different versions have been preserved among the manuscripts collected in Kings College alone. The title "Irenicum" is misleading. Undoubtedly the treatise was partly an expression of the irenical movement which explored the fundamentals on which the different Christian groups could be reconciled and reunited. "Irenicum" was more than irenical, however, for the interesting part of the treatise is the formula on which Newton proposed to make peace. Most Christian denominations would surely have hesitated to accept a statement of faith that equated Christianity with natural religion. The original and pure religion, Newton maintained, was the moral religion which was plain to all men, love of God and love of neighbor. The natural product of human reason, it prevailed among the uncorrupted men of the world's youth. Through the ages, however, mankind continually polluted the pure religion by introducing new and excess articles of belief. As often as men defiled it, new prophets from God were sent to revive and to restore the original worship. Noah was one such prophet, Moses another. Finally Jesus Christ appeared, the last of the prophets sent to purify religion. He added nothing to the true religion except the belief that He arose from the dead and that because of His obedience He can prevail upon God to forgive sinners. The fundamental purpose of Christ's mission was to confirm the true primitive religion. What is the basic creed of Christianity?—to love God with all your heart and to love your neighbor as yourself. Christianity does not differ from the natural religion known to all rational men. Those who would add articles of belief—and here Newton came to the burden of his argument —are corrupters of the pure religion, schismatics dividing the body of the faithful.[20] In effect, Newton's "Irenicum" dispensed with revelation, radically redefined the concepts of atonement

20. The drafts of "Irenicum" are collected in Kings College, Keynes MSS, No. 3; four are printed in Isaac Newton, *Theological Manuscripts*, ed. Herbert McLachlan, Liverpool, 1950; one is printed in Brewster, *Newton*, 2, 526–31.

and redemption, and in sum rejected everything that distinguishes Christianity from natural religion.

A passage from a second manuscript treatise, "A Short Scheme of the True Religion," illustrates the extent to which Newton confounded Christianity with natural reason. In discussing what he considered to be the second of the two duties of religion, love of neighbors or fellow men, he stated that the golden rule was acknowledged by heathens and ought to be the law of all mankind.

> This was the ethics, or good manners, taught the first ages by Noah and his sons in some of their seven precepts, the heathens by Socrates, Cicero, Confucius and other philosophers, the Israelites by Moses and the Prophets, and the Christians more fully by Christ and His apostles. This is that law which the apostle tells you was written in the hearts of the gentiles and by which they were to be judged in the last day. . . . Thus you see there is but one law for all nations, the law of righteousness and charity dictated to the Christians by Christ, to the Jews by Moses, and to all mankind by the light of reason, and by this law all men are to be judged at the last day.[21]

If Newton's basic contention, that the Golden Rule is dictated by the light of reason, is granted, he succeeded in his effort to place Christianity as he understood it on a rational plane, but what was the Christianity that he understood? Through his words blew the chill wind of death for Christian revelation, for Newton equated Christ with reason and pruned Christianity of all supernatural elements.

Despite phrases such as "written in the heart" and "dictated by the light of reason" Newton did not usually think of natural religion as a set of innate principles stamped in the heart of every man. As a virtuoso accustomed to seek knowledge through the observation of nature, he believed that the existence of God and man's duties toward Him were easily deduced from the natural order. In this opinion he associated himself with the main stream of thought on natural religion among the virtuosi, binding scien-

21. Kings College, Keynes MSS, No. 7; printed, with minor errors, in Newton, *Theological Manuscripts*, p. 52.

tific studies closely to the cause of religion. The passage concluding the thirty-first query in the *Optics*, one of the more important revelations of Newton's religious ideas, is apt to be overlooked as a conventional statement of pious purpose. When read in conjunction with Newton's theological manuscripts, it takes on new meaning and helps to point up the intimate connection between his science and his religion. If natural philosophy shall at length be perfected, Newton suggested, the bounds of moral philosophy will also be enlarged. "For so far as we can know by natural philosophy what is the First Cause, what power He has over us, and what benefits we receive from Him, so far our duty towards Him as well as that towards one another will appear to us by the light of nature." [22] No doubt, Newton added, the heathens would have gone further in their moral philosophy had they not blinded themselves by the worship of false gods. Instead of teaching the worship of the sun and the moon and dead heroes, they might have taught us to worship the true Author of our being as their ancestors did under Noah before they corrupted themselves. That is to say, natural religion, true religion, is revealed by natural philosophy. When Newton told Richard Bentley that he wrote the *Principia* with an eye to principles that would prove the existence of God, he was expressing the fundamental creed of his religion. By revealing the Author of nature and through Him the basic duties of love for God and neighbor, natural philosophy places religion upon the footing of reason for which Newton and the virtuosi were looking. Nature lies open before all men; since they have common faculties with which to study nature, all men should reach the same conclusion. Natural religion supplied the object for which all the virtuosi seemed to be searching, a statement of religious fundamentals based on principles that they could test and accept. Newton's only difference was to equate natural religion with the whole of Christianity.

There is a distinction between rejecting Christianity in favor of natural religion and equating Christianity to natural religion. Newton continued to call his religion Christianity. We have already seen how he worked over the doctrine of providence, un-

22. *Optics*, pp. 381–2.

able quite to assimilate it into the mechanical philosophy, yet unwilling to let it go. In much the same spirit he clung to the name "Christianity" which decorated the surface of his natural religion without changing its substance. The divinity of Christ is a key issue; Newton repudiated it. For two centuries following his death his rejection of the Trinity remained obscure until the publication of some of his private papers in the 20th century put the issue beyond doubt. One paper, a set of twelve doctrinal positions, makes reference to "the man Christ Jesus," while the remaining eleven points deepen the distinction between the supreme God and His created servant, Jesus.[23] Other papers have fourteen *Argumenta,* supported by Scriptural passages, proving that the Son is neither coeternal with nor coequal to the Father, seven *Rationes* against the Trinity, and memoranda defining faith in Christ in a manner suggesting that He was the human agent of God.[24] Nevertheless Newton can also be found referring to Christ as "God" and declaring that He redeemed man with His blood. By verbal tricks such as these Newton hoped to maintain his religion, not as a rejection of Christianity but as its true interpretation. He argued, for instance, that usage in the Old Testament applied the name "God" to men who received the word of God directly. Since the titles given to Jesus—such as Messiah, Lamb of God, Son of God, and Son of Man—were derived from the Old Testament, they must be interpreted according to the Old Testament's usage. Thus it is acceptable to refer to Christ as "God," though to worship Him as God would be blasphemy. Newton's treatment of Christ's redemptive role, an argument in which the influence of Locke is evident, attempted in the same way to gloss over the question with a verbal formula employing the traditional terms in a different sense. Christ, he began, was raised from the dead; and in like manner all men will at length be raised by God and judged according to their merits. Because of Christ's obedience God has given Him a kingdom to be selected from the best of men; at Christ's request their sins will be remitted.

23. Kings College, Keynes MSS, No. 8; printed in Brewster, *Newton,* 2, 349–50.
24. All cited, with extracts, in More, *Newton,* pp. 642–3.

> For all men except Christ are sinners and might in justice be punished for their sins; but Christ, by His obedience to God and particularly by His submitting to God's will even to die an ignominious and painful death upon the cross as an example to teach us absolute obedience in all things to God's will, has so far pleased God as to merit of Him a kingdom and that God should forgive the sins of all those whom He shall choose to be His subjects and therefore He is said to have made an atonement for us, and to have satisfied God's wrath and merited our pardon, and to have washed away our sins in His blood and made us kings and priests.[25]

In both cases Newton set up words in place of real things. Christ might be called "God"—but it was only a name. Christ might be called our Redeemer—but the process of redemption was a private transaction between Jesus and God which was wholly removed from the life of the individual Christian struggling with his sins. Behind Newton's use of traditional Christian terms the substance of traditional Christian doctrines had disappeared.

The point at issue is not Newton's sincerity. His intent was not to hide infidelity behind a façade of orthodox phrases. Since the papers in which his religious opinions are found were private manuscripts which he had no intention of publishing, he would have had no reason not to express himself fully. Instead of a hypocrite Newton was a man driven at once by a compulsion to retain the Christian religion and by a revulsion from facets of it that he considered irrational. Hence he tried to explain away the "irrationalities" without destroying the forms. To be sure there was an element of sham in Newton. Heretical, according to the standards of the day, in that he rejected the Trinity, he was fearful of the onus attached to heresy and chose to hide his true beliefs. But heresy is not infidelity. To repudiate the Trinity was one thing, to repudiate Christianity something far greater. While his private papers make it clear that secretly he rejected the Trinity, there is no evidence whatever to suggest that he separated himself from Christianity, even in the privacy of his thoughts, to em-

25. Kings' College, Keynes MSS, No. 3.

brace, say, deism. Indeed it is manifest that Newton considered himself a Christian; and since a man's intention is all-important, he must be accepted as such—a Christian, though a Christian who rationalized his belief. Some of his antitrinitarian papers suggest the key to his attitude. Even in his private writings Newton hesitated to make the final irrevocable statement rejecting the Trinity; he soothed his sensibilities as it were by couching his disavowal in indirect terms. The essay in biblical criticism, "An Historical Account of Two Notable Corruptions of Scripture," is certainly antitrinitarian, but it tries to pose as nothing more than the exposure of two spurious texts used to support the Trinity. Other papers, "Paradoxical Questions Concerning the Morals and Actions of Athanasius and His Followers," and "Queries Regarding the Word Homoousios," adopt the same indirect method. Although some of the manuscripts are more specific, none of them makes a fully direct and explicit statement. The same spirit, the reluctance to surrender established forms, guided Newton's treatment of Christianity as a whole. While he rationalized away the received meaning of many doctrines, he clung to the name Christian, considered himself a Christian, and believed that his interpretation was the true Christianity. The impossibility of choosing finally between different interpretations may be granted, yet how little of the substance that for centuries had been taken as Christianity remained in Newton's formulation! Only the name was there; the traditional religion was gone. When he finished defining Christ and His role for man, Newton had left the natural religion outlined in the "Irenicum" virtually unchanged. The duties of religion were still love of God and love of man, the naturally revealed duties to which Christ added nothing. Fulfillment of these duties was left to the unaided natural man; there was no idea of grace through Christ. The process of salvation lost the aspects of spiritual growth in man and of supernatural aid from God. Newton's discussion of Christ's redemptive role left salvation as the reward for natural virtue. Christ the Redeemer, as defined by Newton, had an artificial and impossible relationship to salvation; had He not been mentioned at all, the rest of Newton's religious structure would not have been disturbed. In a word Newton's Christianity was indistinguishable from natural religion.

Why did Newton reduce Christianity to a bare foundation? The first of his seven *rationes* against the doctrine of the Trinity indicates the answer, not in express words but in the spirit which informs both the statement and the whole corpus of Newton's religious writings. "*Homoousion* is unintelligible," he declared. " 'Twas not understood in the Council of Nice [*sic*] . . . nor ever since. What cannot be understood is no object of belief." [26] Religious mysteries could not survive the drive for complete intelligibility. Given the attitude, Newton's conclusions followed logically. The spirit behind the statement was not Newton's invention. He shared it with the other Christian virtuosi; indeed he inherited it from them. In essence it had been the motivation behind their expositions of natural religion. Initially the demand for intelligibility had been applied in a limited sense. Boyle, for instance, had not wished to deny religious mysteries; his purpose had been to prove that the foundations of Christianity are demonstrable by reason. Altering Newton's phrase, Boyle might have said, "What can be understood can be the foundation of belief." Newton only universalized the search for rationality by insisting that all articles of belief march under the banner of reason. If he went beyond the other virtuosi, at least he went in the direction in which they had been moving.

Worship of the Almighty Creator was what remained of Christianity when the unintelligibles had been lopped off—the Almighty Creator, Boyle's Divine Watchmaker, Ray's Architect of ten thousand species, Newton's First Cause.

> We are therefore to acknowledge one God [says the "Short Scheme of the True Religion"], infinite, eternal, omnipresent, omniscient, and omnipotent, the Creator of all things, most wise, most just, most good, most holy; and to have no other Gods but Him. We must love Him, fear Him, honor Him, trust in Him, pray to Him, give Him thanks, praise Him, hallow His name, obey His commandments, and set times apart

26. Cited in More, *Newton*, p. 642.

> for His service, as we are directed in the third and fourth commandments.[27]

This, he added, is the first and principal part of religion. As natural religion came to dominate Christianity, the figure of the Son, love mediating between God and man, faded before the image of the Father, the powerful and transcendent Creator. There is little in the passage just quoted to suggest a close spiritual relationship between the Almighty and His creature, man. Newton's God was a distant, removed Being Whom man could acknowledge but scarcely enjoy. Significantly enough Newton listed acknowledgment first among the duties. "We are to acknowledge one God . . ." His religious attitude as a whole suggests that the remaining duties were really comprehended in the first. An affair of the head but not of the heart, Newton's religion was an intellectual exercise approximating the study of natural philosophy. Through the observation of nature men could learn of the Creator and of their duties toward Him. But it was all so cold! The warmth of spiritual devotion had dissipated before the chilling rays of intellectual clarity.

Scholarship has preserved a tradition that Newton was a mystic. The tradition has endured into the 20th century, surviving the recent publication of the papers in which Newton revealed himself as a rationalistic antitrinitarian. Recently it has received its most vigorous expression in the final chapter of Professor Andrade's *Newton.*[28] Newton's interest, or reputed interest, in the writings of Jacob Boehme lies at the basis of the contention that he was a mystic. According to the story he read deeply in the German theosophist and copied long extracts from his works. The argument for mysticism is supported by Newton's study of alchemy, and his writings on the prophecies further buttress the case. On closer examination, however, the view of Newton as a mystic is seen to rest on the flimsiest evidence expanded by liberal conjecture, while the solid evidence in his theological manuscripts opposes it squarely. The story of Boehme's influence would

27. Kings College, Keynes MSS, No. 7; printed in Newton, *Theological Manuscripts,* p. 51.

28. E. N. da C. Andrade, *Isaac Newton,* London, 1950.

seem to be unfounded rumor. It has been traced to William Law, who first mentioned Newton's interest in Boehme in the middle of the 18th century, while there is no evidence of the interest dating from Newton's own lifetime. Although the extracts from Boehme continue to be cited, they do not exist among Newton's papers, and no one has been able to locate them. Newton's deep concern with alchemy is beyond question; voluminous alchemical treatises copied out in his own hand compose a large segment of his papers. Once again, however, there is no evidence that he was attracted to the mystical side of alchemy, while there are good indications that his interest may have concentrated on whatever scientific information alchemy contained. Evidently Newton tried to discover the force that holds atoms together in solid matter. He realized that gravity cannot account for the different intensities of chemical reactions, which we might attribute to chemical affinities today. Although he was never able to arrive at an established conclusion on the subject, he conjectured that another force must be at work. Some indications suggest that Newton looked into alchemy for an approach to this force. His short essay on the nature of acids, for example, and the thirty-first query of the *Optics* use the vocabulary of alchemy in discussing the interatomic force. His theory of the ether, which was an intimate part of his conjectures about the atomic force, bears a remarkable resemblance to some of the alchemical treatises in its picture of nature in constant transmutation. These examples suggest the influence of alchemy, but its esoteric and mystical element is absent. Meanwhile the evidence of Newton's religious writings makes it difficult to believe that the mysticism associated with alchemy could have attracted him. His *Observations upon the Prophecies* serve the argument for mysticism no better. Essays in dry historical symbolism, they proceed on the assumption that the prophecies were written in a definite code of prophetical language. The key can be discovered without the aid of special divine light and used to decipher them like any coded message. They can be used only to interpret the past. Since Newton explicitly spurned the notion that the prophecies can enable man to foretell the future, he did not use them to predict any millennium or utopia. By comparing the prophecies with recorded history, which of course still lay in the future as far as the authors of the prophecies were concerned, he merely sought to demonstrate God's governance of the world through His plan for human society. In much the same way the study of natural philosophy was supposed to reveal His dominion over matter. In short, Newton's *Observations upon the Prophecies* are not mystical writings; they are historical and interpretative essays. All of the evidence for mysticism falls to the ground, while the indisputable evidence of Newton's religious opinions, found in his own manuscripts, points in the other direction. His own words in a vehement assertion of rationalism seem meant expressly to reject the notion that he was a mystic. "It is the temper of the hot and superstitious part of mankind," the scientist insisted with some warmth, "in matters of religion ever to be fond of mysteries, and for that reason to like best what they understand least." For his part, he declared, he loved to take up with what he could best understand.[29] How can the man who uttered these sentiments be made into a mystic? How can the man who wrote and rewrote the essay on natural religion, "Irenicum," and the equally rationalistic "Short Scheme of the True Religion" be made into a mystic? All of the established evidence testifies that Newton was a religious rationalist who strove ever to reach a satisfactory intellectual definition of Christianity but remained blind to the mystic's spiritual communion with the divine.

The lack of spiritual depth in Newton's religious writings can be seen in his treatment of the two duties that he considered fundamental to religion, love of God and love of neighbor. Surely here was the opportunity to introduce the life of the spirit into the bare framework of definitions; yet in Newton's hands the acts of love transformed themselves into a set of negative "Thou-shalt-not's." A communicant of the true religion, according to one draft of the "Irenicum," is required to repent his breaches of the two commandments, which is called forsaking the world and the devil.

> To forsake the devil [Newton continued, defining the fundamental duties of religion] is to forsake all false Gods, the

29. Isaac Newton, *Opera quae Exstant Omnia,* ed. Samuel Horsley (5 vols. London, 1779–85), 5, 529–30.

> worship of which is a breach in the first and great commandment. To forsake the world is to abandon the love of the things of this world, called by the Apostle John the lust of the flesh, the lust of the eye, and the pride of life, that is, the inordinate desire of women, riches, and honor, or effeminacy, covetousness, and ambition, which are the root of all evil against our neighbor and the fountains of uncharitableness.[30]

Nowhere did Newton offer a positive definition or description of the fundamental duties. Instead of a spiritual act love became a moral code proscribing certain activities; it became susceptible of intellectual definition; it became intelligible. His handling of the basic duties of love epitomizes the end product of Newton's religious thinking. In his drive for a rationally demonstrable religion he excluded the spiritual elements of Christianity. All that survived was the acknowledgment of God the Creator and the recognition of His moral law. Worship was equated with obedience. Newton's writings have the flavor of a man who reduced all religious questions to the intellectual or semantic plane. He repudiated the Trinity and the divinity of Christ because he thought that they were physically impossible. Physical possibility, however, is not the only consideration involved in the divinity of Christ; the sinfulness of man, his need for redemption, his inability to save himself without divine aid—these matters also bear upon the subject. A man's response to the spiritual experience behind religious doctrines will help to shape his formalized theology. Newton simply ignored the spiritual questions. Christianity was to him a matter of doctrines, or rational formulae. Thus he wrote endlessly, defining the true religion, but never did he prostrate himself before his God.

With Newton, Christianity's history among the virtuosi reached a turning point. He combined and consolidated the series of modifications in the Christian tradition made by the virtuosi. Each modification was relatively minor in itself, and none of them constituted a break with Christianity, yet their cumulative effect as seen in Newton's statement of Christianity attenuated the Christian tradition to the verge of disintegration. Further change,

30. Kings College, Keynes MSS, No. 3.

within the intellectual framework prevailing in the late 17th century, could proceed only by repudiating Christianity. The change in religious thought did not progress by contradiction and open conflict; more than controversy, reinterpretation took place within the bosom of the Christian religion. The modifying influences, among which natural science was only one element in a complex of intellectual and social forces, worked beneath a cover of Christian piety which obscured their ultimate implications. The reverence that nature inspired in the virtuosi upheld their religious sincerity, maintaining the structure of Christianity within which they worked. Within those limits they gradually elaborated an interpretation of providence that limited God's role in creation to sustaining the universal order. To demonstrate the rational foundation of Christianity they reasoned from natural philosophy to natural religion and finally concluded that the propositions of natural religion were more important than the doctrines revealed in the Bible. Against a background of sectarian controversy over the Scriptural message the implication always present, that the "fundamental" truths revealed in nature are the real essentials of religion, grew to explicit recognition of natural religion's superior importance. The declining position of the Bible was further hastened by the growing confidence in human reason, which led to preference for clear and demonstrable propositions and suspicion of revealed mysteries. Taken together, these tendencies meant that the avowed relative positions of natural science and religion had in reality been reversed; Christianity was being judged by its old handmaiden and adjusted to conform to her dicta. The world view of natural science had achieved predominance. Science now supplied the criteria of truth. When Newton consolidated these elements into his statement of Christianity, the cover of Christian piety still sanctified his labors; but traditional Christianity trembled on the brink of dissolution.

Little separated Newton's religion from the 18th century's religion of reason—only the name "Christianity" and an attitude which the name implied. The virtuosi had taken up natural religion originally in defense of Christianity, and this attitude still remained dominant in Newton. In removing the fragments of irrationality he was saving Christianity from itself and defending it from skepticism. The virtuosi's concentration on natural religion meant that they treated only those aspects of Christianity to which rational proofs might apply, while they ignored the spiritual needs to which Christianity had ministered through the centuries. In defending Christianity in this manner, they prepared the ground for the deists of the Enlightenment—the mechanical universe run by immutable natural laws, the transcendent God removed and separated from His creation, the moral law which took the place of spiritual worship, the rational man able to discover the true religion without the aid of special revelation. The religion of reason grew to maturity in the tradition that Newton completed, but the conviction that it was Christianity still remained. Newton did not look upon himself as a skeptic or an infidel. He thought that he had the real Christianity, safe at last behind its wall of rational demonstration. Unhappy thought! Change only the attitude, remove the reverence for Christianity that the virtuosi maintained, in a word move only from the religious 17th century into the doubting 18th, and deism, the religion of reason, steps full grown from the writings of the Christian virtuosi.

We might better say that the difference in attitude between the virtuoso and the *philosophe* was not reverence opposed to skepticism but rather uncertainty opposed to open doubt. Despite the natural piety of the virtuosi, the skepticism of the Enlightenment was already present in embryo among them. To be sure, their piety kept it in check, but they were unable fully to banish it. What else can explain the countless dissertations on natural religion, each proving conclusively that the fundamentals of Christianity are rationally sound? They wrote to refute atheism, but where were the atheists? The virtuosi nourished the atheists within their own minds. Atheism was the vague feeling of uncertainty which their studies had raised, not uncertainty of their own convictions so much as uncertainty of the ultimate conclusions that might lie hidden in the principles of natural science. With wonderful certainty and assurance each virtuoso proved the existence of God from the creation; yet repeated too often, the assurance acquired an odor of insecurity. With Newton the insecurity was growing toward open fright. The creation pointed infallibly to the First Cause, but was Christianity itself entirely rational? Could it stand the test of reason? Did it not need to be purged before it could be safe? Newton wrote a paper to prove to himself that every doctrine of the true Christianity was rational and reasonable. Somehow it was not quite right. He revised it, wrote it again, wrote it a fourth time, and then a fifth. Still it was not quite right. Perhaps if he tried once more, he could reach the perfect statement, the exact definition which would reconcile Christianity with reason forever and restore certainty to religion. That picture of Newton in his old age writing and revising his statement on religion is the symbol of the insecurity that goaded the virtuosi as they sought a foundation for certainty. But certainty there was not to be. Following the birth of modern science the age of unshaken faith was lost to western man.

Bibliographical Essay

THE FIRST EFFORTS of modern historiography to assess the problems of science and religion in the 17th century were made in the latter half of the 19th. Three historians who examined the subject stood on the side of science in the Darwinian controversy. Since they wished to emphasize the harm that had been done in the past when scientific discoveries had been opposed in the name of religion, their books were to some degree tracts for the times as well as historiographical studies. Moreover, their works are pervaded with the smug satisfaction of an age that believed in progress and looked upon itself as the worthy product of the centuries' toil. The civilization of the 19th century was good; anything that had opposed its origins must therefore have been bad. W. E. H. Lecky, *History of the Rise and Influence of the Spirit of Rationalism in Europe* (2 vols. London, 1865), pictures the rise of rationalism as the triumph of civilization over barbarism. The triumph of civilization, in turn, was marked by the growth of science and the decline of dogmatic theology. To be sure, Lecky does not equate science with rationalism. Although he considers natural science to be an important element of rationalism, it is still only one element in a complex of ideas that includes disbelief in the miraculous, a refined concept of God as a rational being instead of an arbitrary ruler, the substitution of moral consciousness for dogmatic theology, the secularization of politics, and the Baconian belief in the Kingdom of Man. Lecky believes that science was antithetical to the theology that dominated thought before the rise of rationalism in the 17th century; it demonstrated that nature is an ordered cosmos governed by law from which the miracles of theology are excluded. Since Lecky does not equate theology with religion, he feels that true religion has been promoted by the decline of theology; in the end science has helped religion by purging it of crude obscurities. John W. Draper, *History of the Conflict between Religion and Science* (New York, 1875), carries the idea of conflict to the extreme, maintaining that the Christian church sternly repressed scientific thought until in the 17th century the pressure of science was too great to be contained. An expression at once of thinly veiled atheism and of hatred for Catholicism, Draper's book contends that from the time of its birth Christianity opposed natural science. By setting up superstition, blind faith, and reliance on the Bible, the Christian church destroyed ancient science, precipitated the Dark Ages, and suppressed enlightenment until the explosion in the 17th century. In his eyes the scientists of the period were the champions of truth against intolerant superstition. Although Andrew D. White, *A History of the Warfare of Science with Theology in Christendom* (2 vols. New York, 1896), adopts a more moderate tone, its theme is adequately summarized in its title. He believes that science had to struggle during its growth, both in the 17th century and in other ages, against the superstition that insisted on the literal meaning of the Bible. White's book is the outcome of his battle with aroused orthodoxy, which condemned freedom of scientific work at Cornell University. Since he wishes to stress the damage both to science and to religion that has followed from every attempt of Christianity to dictate scientific truth, he restricts his study to the controversies and largely to the opinions of theologians. In so doing, he leaves the impression that theology has blindly opposed every advance toward scientific truth and that every scientist has been drenched in a torrent of theological abuse. With Lecky, White draws a distinction between theology and religion; he insists that the ultimate effect of science has benefited religion by cleansing it of primitive and mistaken beliefs. A more recent study, Paul Hazard, *La Crise de la conscience Européenne* (3 vols. Paris, 1934), repeats the emphasis on conflict between science and religion. Although Hazard deals with much more than science, he considers it as one aspect of the general questioning and rejection of ancient standards during the period of "crisis," 1680–1715.

Not only has Christianity become more intellectually respectable in the 20th century, but the confident temper of the 19th has faded in many circles and with it faith in the benevolence of natural science. These changes have been reflected in some of the scholarship dealing with the 17th century. One school has continued to emphasize the incompatibility of science with Christianity, but it investigates the ultimate implications of science instead of immediate quarrels with theology. It is also more concerned with the harmful effects of the scientific outlook on Christianity than with the opposition of Christian theology to science. E. A. Burtt, *The Metaphysical Foundations of Modern Physical Science* (London, 1925), argues from the point of view of outraged Christian humanism. Burtt maintains that the corrupted form of Cartesian dualism that came down to the modern world through Newton's publications, gaining general acceptance from its alliance with Newton's name, denies any reality to the values and aspirations of mankind. Newton's conception of nature atrophied religion, picturing God in the ridiculous role of cosmic plumber; while the following century proved that even the plumbing functions of the Deity are unnecessary. God was portrayed as an abstract original cause, distant and removed from man. Since most men are incurably anthropomorphic in their conception of God, the scientific world view undercut religion and prepared its destruction. When Richard F. Jones, *Ancients and Moderns* (St. Louis, 1936), touches on science and religion, he repeats Burtt's attitude and approach. Jones finds influences inimical to Christianity not only in the naturalism of science but also in its "materialism," by which he means its Baconian utilitarianism. In the same general school Basil Willey's *Seventeenth Century Background* (London, 1934) surveys the ultimate implications of science from a slightly different angle. The major intellectual problem of the 17th century, according to Willey, was the winnowing out of truth from falsehood, superstition, and fable. All of the century's thinkers were concerned to discover the real and the true. Ignoring the questions of medieval philosophy, they demanded an explanation of how things are caused instead of why; that is, they demanded knowledge that would bestow power to control nature. The real ceased to be the hierarchy of metaphysical orders and the ultimate causes and purposes of things and became the hard cold world of geometric matter. Since clear and distinct ideas were the criteria of truth, imagination was distrusted and rejected, to the detriment both of poetry and of religion. Since the faculty of the mind that finds truth in concrete human experience was suspect, religion became a matter of intellectual demonstrations—abstract propositions instead of living experience. Thus in its ultimate effect the intellectual development of the 17th century undermined vital religion.

Although Burtt, Jones, and Willey contend that early modern science and Christianity were incompatible, they also conclude that the scientists of the period did not consider the two to be incompatible. Both Burtt and Jones mention the religious value that the virtuosi found in their work. Willey pursues the idea at greater length. In *The Sevententh Century Background;* in an essay, "The Touch of Cold Philosophy" (in the *festschrift* for Richard F. Jones, *The Seventeenth Century*, Stanford, 1951); and in the opening chapters of *The Eighteenth Century Background* (London, 1940) he points out that the scientists reconciled belief in God and acceptance of natural science by picturing God as the Creator of the ordered cosmos. In the chapter devoted to the 17th century in his recent volume, *Christianity, Past and Present* (Cambridge, 1952), Willey indicates both the religious inspiration that the virtuosi found in contemplating the order and laws of nature and the application of their conclusions to the proof of God's existence when Christianity seemed to be challenged by Hobbesian materialism. In a chapter devoted to the influence of science on religion G. R. Cragg, *From Puritanism to the Age of Reason* (Cambridge, 1950), also elaborates upon the efforts of the virtuosi to find religious significance in their work.

Another aspect of the relations of science and religion in the 17th century is explored in a series of works beginning with Alfred North Whitehead, *Science and the Modern World*, New York, 1925. Whitehead argues that the concept of an ordered cosmos, which is the foundation of modern science, was the legacy of the Christian philosophy of the Middle Ages. R. G. Collingwood dilates upon the suggestion in his *Idea of Nature*, Oxford, 1945. Collingwood holds that the Renaissance conception of nature, by which he means the mechanical conception of nature worked out in the 16th and 17th centuries, pictured nature as an intelligent order organized by a mind external to it. The basic notion derived from the Christian concept of creation and from practical experience with machinery. Charles E. Raven, *Natural*

Religion and Christian Theology (Cambridge, Cambridge University Press, 1953), gives the idea a different meaning. Regarding the medieval dualism of body and spirit as a repudiation of the Greek and Hebrew traditions, Raven equates true Christianity with the conception of nature as the continually creative manifestation of the immanent spirit of God. In his opinion this conception was the driving force behind what he considers the most important aspect of 17th-century science, the tradition of naturalists culminating in John Ray. Raven's *John Ray, Naturalist* (Cambridge, 1942) may be looked upon as a case history of the influence of Christianity upon the development of modern science. The idea that Christianity influenced the growth of modern science also lies behind the investigations of Puritanism and science. The literature on this subject is surveyed in footnote 1 of the first chapter.

The study of individual virtuosi has provided a different approach to the science and religion in the 17th century. In general, the biographies of virtuosi stress their belief in the harmony of natural science and Christianity. Those studies which are most penetrating on the subject include Louis T. More, *Isaac Newton, a Biography*, New York, 1934; Louis T. More, *The Life and Works of the Honorable Robert Boyle*, New York, 1944; Charles E. Raven, *John Ray, Naturalist*, Cambridge, 1942; Percy H. Osmond, *Isaac Barrow. His Life and Times*, London, 1944; and H. R. Fox Bourne, *The Life of John Locke*, 2 vols. London, 1876. Easily the most detailed investigation of science and religion among the virtuosi is Mitchell S. Fisher, *Robert Boyle, Devout Naturalist*, Philadelphia, 1945. As the title suggests, Fisher emphasizes Boyle's reconciliation of science and religion. He interprets Boyle as a religious conservative who followed the Church without question in matters of faith, and compromised his science instead of his religion whenever the possibility of conflict arose. Moody E. Prior, "Joseph Glanvill, Witchcraft, and Seventeenth-century Science," *Modern Philology, 30* (1932–33), 167–93, argues that Glanvill's defense of witchcraft was part of a general attempt by the virtuosi to prove that mechanistic science did not lead to atheism. John Locke's religious opinions have been the subject of studies too numerous to be mentioned here. Among the most informative is Sterling P. Lamprecht, *The Moral and Political Philosophy of John Locke,* New York, 1918. Herbert McLachlan, *The Religious Opinions of Milton, Locke and Newton* (Manchester, 1941), is a Unitarian's examination of the opinions of the three men on the Trinity. Stephen Hobhouse, "Isaac Newton and Jacob Boehme," *Philosophia* (Belgrade, Yugoslavia), *2* (1937), reprinted as Appendix Four in Hobhouse's volume, *Selected Mystical Writings of William Law* (London, 1948), is an important contribution to the comprehension of Newton's religious thought. Hobhouse effectively disposes of the argument that Boehme's mysticism exercised a great influence on Newton's religious thought.

The classic treatment of the conception of nature on which natural science rested in the 17th century is Edwin A. Burtt, *The Metaphysical Foundations of Modern Physical Science,* London, 1925. Edward W. Strong, *Procedures and Metaphysics* (Berkeley, 1936), attempts to revise Burtt by contending that the mathematical method of analyzing nature preceded the metaphysics that pictured nature as a mathematical order. The section in R. G. Collingwood, *The Idea of Nature* (Oxford, 1945), that deals with the 17th century is a briefer but penetrating analysis of the virtuosi's conception of nature. Marie Boas studies its growth in "The Establishment of the Mechanical Philosophy," *Osiris, 10* (1952), 412–541. Arthur O. Lovejoy, *The Great Chain of Being* (Cambridge, Mass., 1936), traces the history of the concept of the chain of being from ancient Greece to the 19th century, pointing out its importance for the scientific thinkers of the 17th century. Among the many histories of science, two of the most informative about science in the 17th century are Herbert Butterfield, *The Origins of Modern Science* (London, 1950), which deals mostly with the 17th century, and Abraham Wolf, *A History of Science, Technology, and Philosophy in the 16th & 17th Centuries* (London, 1935), which surveys the development of each field of science in detail.

Recently science as a social phenomenon has been studied extensively. Its relations with society in general, with economics, and with technology are discussed with penetration by G. N. Clark, *Science and Social Welfare in the Age of Newton,* Oxford, 1937. Clark's work replies to the essay by B. Hessen, "The Social and Economic Roots of Newton's Principia," *Science at the Cross Roads,* London, 1931. Hessen gives a Marxian interpretation to the origin of modern science, while Clark specifically argues against the Marxian theory. Robert K. Merton examines the social and religious bases of the rise of science in his pioneering study, "Science, Technology and Society in Seventeenth Century England," *Osiris, 4* (1938), 360–632. Harcourt Brown, "The Utilitarian Motive in the Age of Descartes," *Annals of Science, 1* (1936), 182–92, limits its discussion to France; while the article by Walter E. Houghton, Jr., "The History of Trades: Its Relation to Seventeenth Century Thought," *Journal of the History of Ideas,* 2 (1941), 33–60, examines an aspect of the utilitarian movement in England.

A survey of the development of theology in England during the 17th century can be found in John Hunt, *Religious Thought in England from the Reformation to the End of the Last Century,* 3 vols. London, 1870–73. W. Fraser Mitchell, *English Pulpit Oratory from Andrewes to Tillotson* (London, 1932), makes the changing structure of sermons reveal major developments in English religious thought during the 17th century. H. R. McAdoo, *The Structure of Caroline Moral Theology* (London, 1949), is a solid piece of work. A collection of extracts from 17th-century Anglicans, *Anglicanism,* ed. Paul Elmer More and Frank L. Cross (London, 1935), is introduced by two good essays on the dominant faith in England. William Haller, *The Rise of Puritanism* (New York, 1938), is the best study of the Puritans during the early 17th century. G. R. Cragg, *From Puritanism to the Age of Reason* (Cambridge, 1950), presents an intelligent discussion of changing religious ideals during the latter half of the century. Attempts to meet the problems posed by religious controversy is the central theme of John Tulloch, *Rational Theology and Christian Philosophy in England in the Seventeenth Century,* 2 vols. London, 1872. Tulloch deals both with the rationalistic theologians of the irenical movement preceding the Civil War and with the Cambridge Platonists. His work on the latter has been superseded by Frederick J. Powicke, *The Cambridge Platonists,* Cambridge, Mass., 1926. The title of Samuel L. Bethell's book, *The Cultural Revolution of the Seventeenth Century* (London, 1951), masks a penetrating discussion of faith and reason. A different aspect of 17th-century religious thought is presented by Ernest Lee Tuveson, *Millennium and Utopia,* Berkeley and Los Angeles, 1949. Tuveson shows how interpretations of the prophecies changed from pessimism to optimism during the 17th century until they became forerunners of the 18th century's idea of progress.

The works of the virtuosi themselves have been cited in the footnotes. Among them Robert Hooke's works, especially *Micrographia* (reprinted as Vol. 13 of R. W. T. Gunther, *Early Science in Oxford,* 14 vols. London and Oxford, 1920–45), present possibly the best insight into the virtuosi's response to nature. The works of Robert Boyle contain the best statements of the virtuosi's conception of nature and of the religious significance that scientific investigation held for them. Probably his most revealing treatises are *The Usefulness of Experimental Philosophy: A Free Inquiry into the Vulgarly Received Notion of Nature,* and *The Christian Virtuoso,* all of which are found in *The Works of the Honourable Robert Boyle,* ed. Thomas Birch, 6 vols. London, 1772. John Ray's *Wisdom of God Manifested in the Works of the Creation* (London, 1691), is another of the important works. Newton's best statements on the universe and God's relation to it are found in the General Scholium appended to *The Mathematical Principles of Natural Philosophy,* trans. Andrew Motte, 2 vols. London, 1729 (or any edition except the first, in which the General Scholium did not appear), and in the Queries attached to the *Optics,* 3d ed. London, 1721 (or any edition except the first, which did not contain the full set of Queries). His ideas on natural religion are found in the volume of his *Theological Manuscripts,* ed. Herbert McLachlan, Liverpool, 1950.

Supplemental Bibliography

THE PAST TWENTY YEARS have contributed fundamental studies to the comprehension of the scientific revolution. The most influential of these have been the works of Alexandre Koyré. His *Études galiléennes*, (Paris, 1939), prior obviously to the last twenty years, has done more than any other work to shape the understanding of science in the seventeenth century. He has also produced *From the Closed World to the Infinite Universe*, (Baltimore, 1957) and innumerable articles, the most important of which are collected in *Metaphysics and Measurement: Essays in the Scientific Revolution*, (Cambridge, Mass., 1968) and *Newtonian Studies*, (Cambridge, Mass., 1965). A. R. Hall has published two important general histories of science during the period, *The Scientific Revolution*, (London, 1954) and *From Galileo to Newton*, (New York, 1963). I have written a briefer book, *The Construction of Modern Science*, (New York, 1971). I. B. Cohen, *The Birth of a New Physics*, (Garden City, 1960), restricts itself to the central strand of 17th century science. Thomas Kuhn, *The Structure of Scientific Revolutions*, (Chicago, 1962), while not specifically about science in the seventeenth century, includes it in an important synthesis on the nature of scientific change. Two general histories of scientific thought include illuminating discussions of the seventeenth century—E. J. Dijksterhuis, *The Mechanization of the World Picture*, trans. C. Dikshoorn, (Oxford, 1961) concludes with the 17th century, and Charles Gillispie, *The Edge of Objectivity*, (Princeton, 1960) begins with the scientific revolution. Robert Kargon, *The English Atomists from Hariot to Newton* (Oxford, 1966) studies one current of natural philosophy that was intimately associated with the theme of my book. Recently, the Hermetic tradition has become the center of renewed attention. The most important books on it in English are Francis Yates, *Giordano Bruno and the Hermetic Tradition*, (Chicago, 1964) and Allen Debus, *The English Paracelsians*, (London, 1965).

Puritanism and science, a question raised by Merton and others, has been a subject of continuing interest. Christopher Hill, *Intellectual Origins of the English Revolution*, (Oxford, 1965) significantly expands the thesis, and a series of articles by Hill, Hugh F. Kearney, and Theodore K. Rabb in *Past and Present* in 1964-66 explore it in detail. Lewis S. Feuer, *The Scientific Intellectual*, (New York, 1963) proposes a diametrically opposed thesis, and P. M.. Rattansi, "Paracelsus and the Puritan Revolution," *Ambix*, 11 (1963), 24-32, alters the whole discussion by pointing out that the "science" associated with the Puritans was the Hermetic tradition.

Scientific organizations in the seventeenth century have continued to be studied. A recent examination of the origins of the Royal Society by Margery Purver, *The Royal Society: Concept and Creation*, (London, 1967), stresses the role of Baconianism. *The Correspondence of Henry Oldenburg*, eds. A. R. and Marie Boas Hall, 8 vols. continuing, (Madison, 1965-) provides a major new source for the Royal Society. George N. Clark, *A History of the Royal College of Physicisans of London*, (Oxford, 1964), the first volume of a projected two, covers the seventeenth century. Charles Webster has written two important articles related to the College of Physicians—"English Medical Reformers of the Puritan Revolution: a Background to the 'Society of Chymical Physitans'," *Ambix*, 14 (1967), 16-41; and "The College of Physicians: 'Solomon's House' in Commonwealth England," *Bulletin of the History of Medicine*, 41 (1967), 393-412.

Many of the individual scientists who receive major attention in this book have been the object of continued study. Barbara Shapiro, *John Wilkins*, (Berkeley, 1969) and Margaret 'Espinasse, *Robert Hooke*, (Berkeley, 1956), are the only modern biographies of the two men. Marie Boas, *Robert Boyle and Seventeenth Century Chemistry*, (Cambridge, 1958) replaces all earlier works on Boyle. Above all, Isaac Newton has become of object of a veritable industry. Several important publications of his papers have appeared and are appearing, all with important introductory essays—A. R. and Marie Boas Hall, eds. *Unpublished Scientific Papers of Isaac Newton*, (Cambidge, 1962); I. B. Cohen, ed., *Isaac Newton's Papers & Letters on Natural Philosophy and Re-*

lated Documents, (Cambridge, Mass., 1958); John Herivel, *The Background to Newton's Principia*, (Oxford, 1965); H. W. Turnbull and J. F. Scott, eds., *The Correspondence of Isaac Newton*, 4 vols. continuing, (Cambridge, 1959 continuing); D. T. Whiteside, ed., *The Mathematical Papers of Isaac Newton*, 4 vols. continuing, (Cambridge, 1967 continuing); I. B. Cohen, *Introduction to Newton's Principia*, (Cambridge, 1971), vol. 1 of a new edition of the *Principia* with variant readings. Cohen, the Halls, Herivel, and Whiteside are also the authors of articles on Newton too numerous to be listed here but referred to in their publications above. The *Texas Quarterly* for 1967 contains one issue devoted to the proceedings of the conference on Newton held at the University of Texas in 1966; it contains the best convenient summary of the present state of Newtonian studies. Among the incredible number of articles on Newton, I shall call attention only to a few that help to illuminate his religious thought—two by J. E. McGuire, "Body and Void in Newton's De Mundi Systemate: Some New Sources," *Archive for History of Exact Sciences*, 3 (1966), 206-48; "Force, Active Principles and Newton's Invisible Realm," *Ambix*, 15 (1968), 154-208; J. E. McGuire and P. M. Rattansi, "Newton and the 'Pipes of Pan'," *Notes and Records of the Royal Society*, 21 (1966), 108-43; David Kubrin, "Newton and the Cyclical Cosmos: Providence and the Mechanical Philosophy," *Journal of the History of Ideas*, 28 (1967), 325-46. Finally, I shall mention my own recent book, *Force in Newton's Physics*, (London, 1971).

Index

Selected Ann Arbor Paperbacks
Works of enduring merit